Barron's Regents Exams and Answers

Chemistry
The Physical Setting

ALBERT S. TARENDASH
Assistant Principal—Supervision (Retired), Department of Chemistry and Physics
Stuyvesant High School, New York, New York

Chemistry/Physics Faculty
The Frisch School, Paramus, New Jersey

Barron's Educational Series, Inc.

All inquiries should be addressed to:
Barron's Educational Series, Inc.
250 Wireless Boulevard
Hauppauge, New York 11788
www.barronseduc.com

ISBN-13: 978-0-8120-3163-8
ISBN-10: 0-8120-3163-6
ISSN 0147-7374

PRINTED IN CANADA
9 8 7 6 5 4 3 2 1

Contents

Regents Examinations, Answers and Self-Analysis Charts   93

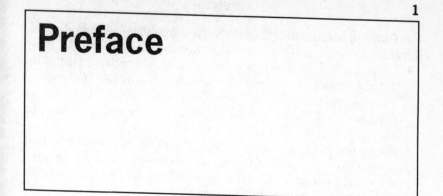

Preface

A HELPFUL WORD TO THE STUDENT:

As you are aware, the purpose of this book is to help you review for the New York State Regents examination in chemistry. You can also use it effectively to prepare for classroom, midterm, and final examinations. The book contains a number of special sections and other features designed to aid you in achieving good grades on your examinations. Included are the following:

- *How to Use This Book.* Your journey begins here. This section explains how to use the *entire* book as an effective test-preparation tool. It provides a method of identifying those areas you have mastered and those that will require additional work.

- *Test-taking Techniques.* In this section, you will learn *how* to take the Regents examination effectively. Included are a list of materials you will need to bring to the examination, directions on how to fill out your answer sheet, how to read the questions, and how to deal with difficult items.

- *What to Expect on the Regents Examination in Chemistry.* This section explains the format, content, and grading of the examination. In addition, it provides some suggestions on how to approach the extended-response questions found on Part *C* of the examination.

- *New York State Regents Chemistry Core: Topic Outline and Question Index.* The topic outline provides a detailed description of the Regents Chemistry Core, and the question index keys the Regents examination questions that appear in this book to the topic outline. The index will help you find questions on similar topics and will provide information about the areas that are stressed most heavily on the exam.

- *Glossary of Important Terms.* The glossary will help you understand the technical terms that may appear on the examination.
- *Using the Reference Tables for Chemistry.* A significant part of the Regents examination requires you to be able to find and use the information contained in these reference tables. This section describes each table and the types of information that may be obtained from it.
- *Recent Regents Examinations.* These questions will provide the bulk of your study and review. When you have answered *all* of the questions that appear in this book, you will be able to approach the Regents examination with confidence!
- *Explanation of Answers and Self-Analysis Chart for Each Examination.* The detailed answer explanations will help you to *understand* why certain choices are correct, and others are incorrect. The self-analysis chart will pinpoint your strengths and weaknesses on each practice examination you take. It will also serve as a tool to help you define those areas in Part II that you *probably* should select and those that you *probably* should avoid.

This book was written to provide you with the basic tools you need to perform well on the Regents Chemistry examination. If you follow its advice and prepare diligently, you will achieve your goal.

Best wishes for success!

Albert S. Tarendash

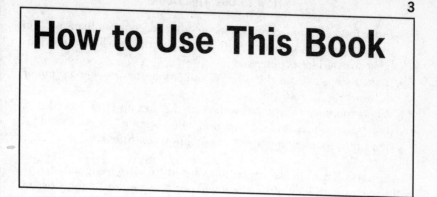

How to Use This Book

1. Read the section entitled *Test-taking Techniques* to learn how to prepare properly for an examination and how to take it with maximum efficiency.
2. Read the section entitled *What to Expect on the Regents Examination in Chemistry* to familiarize yourself with the structure and contents of this examination.
3. Read the section entitled *Using the Reference Tables for Chemistry* to familiarize yourself with the contents and use of these tables.
   Note: On occasion, the New York State Education Department may change the content or format of the Regents examination and/or the reference tables. Your classroom teacher is your best source of information about such changes.
4. Read the section entitled *Using the Equations to Solve Chemistry Problems.*
5. Take the first Regents examination in this book, answering all of the questions.
6. Refer to the *Glossary of Important Terms* to learn the meanings of words and terms you do not understand.
7. Check your answers, and then complete the self-analysis chart at the end of the examination to pinpoint your strengths and weaknesses.
8. Read the detailed explanation of *all of the questions,* paying closest attention to the questions you answered incorrectly. Occasionally, the *wrong choices* are explained, and these explanations may help you understand why you chose an incorrect answer.
9. When you have determined your areas of weakness, refer to the *New York State Regents Chemistry Core: Topic Outline and Question*

Index to locate similar questions on other recent examinations. (You can also use this outline and index to determine which areas have been stressed in recent years.)

10. Repeat steps 5–9 for the other examinations, *with the exception of the most recent test.*
11. When you have completed your studying, but no more than 1 or 2 days before the actual examination, take the most recent examination in this book *under strict examination conditions.*

After you have checked your answers to this last examination, you will have a rough idea of how you will perform on the Regents examination you will take.

Test-Taking Techniques

HELPFUL HINTS

The following pages contain 8 tips to help you achieve a good grade on the Chemistry Regents examination.

TIP 1

Be confident and prepared.

SUGGESTIONS

- Review previous tests.
- Use a clock or watch and take the most recent exam at home under examination conditions (i.e., don't have the radio or television on.)
- Get a review book. (One useful book is Barron's *Let's Review: Chemistry*.)
- Talk over the answers to questions on these tests with someone else, such as another student in your class or someone at home.
- Finish all your homework assignments.
- Look over classroom exams that your teacher gave during the term.
- Take class notes carefully.
- Practice good study habits.
- Know that there are answers for every question.
- Be aware that the people who made up the Regents examination want you to succeed.

- Remember that thousands of students over the last few years have taken and passed a Chemistry Regents. You can pass too!
- Complete your study and review at least one day before the examination. Last-minute cramming does not help and may hurt your performance.
- On the night prior to the exam day, lay out all the things you will need, such as clothing, pens, and admission cards.
- Go to bed early; eat wisely.
- Bring the required materials to the examination. This generally means a pen, two sharpened pencils, and a good quality eraser. If your school does not supply a calculator, be certain to bring one to the examination. Some schools require a signed Regents admission card for identification. Good advice: Assume your school will *not* supply you with any materials!
- Once you are in the exam room, arrange things, get comfortable, be relaxed, attend to personal needs (the bathroom).
- Keep your eyes on your own paper; do not let them wander over to anyone else's paper.
- Be polite in making any reasonable demands of the exam-room proctor, such as changing your seat or having window shades raised or lowered.

TIP 2

Read test instructions carefully.

SUGGESTIONS

- Be familiar with the format of the examination.
- Know how the test will be graded.
- If your school supplies an electronic scoring sheet, be certain you are familiar with the additional directions for recording and changing answers.
- If you decide to change an answer, be certain that you erase your original response completely.
- Any stray marks on your answer sheet should be erased completely.

- Be familiar with the directions for Parts *B* and *C* questions. Answer each question completely. Explanations must be written as *whole sentences* and substitutions into equations *must* include units. Be certain that your answers are clearly labeled and well organized. Place a box around numerical answers. Be neat!
- Ask for assistance from the exam-room proctor if you do not understand the directions.

TIP 3

Read each question carefully and read each choice before you record your answer.

SUGGESTIONS
- Be sure you understand *what* the question is asking.
- Try to recognize information that is *given* in the question.
- Will a chemistry reference table help you find the answer to the question?
- Some choices may look appealing yet will be incorrect.
- Try to eliminate those choices that are *obviously* incorrect.

TIP 4

Budget your test time (3 hours).

SUGGESTIONS
- Bring a watch or clock to the test.
- The Regents examination is designed to be completed in $1\frac{1}{2}$ to 2 hours.
- If you are absolutely uncertain of the answer to a question, mark your question booklet and move on to the next question.
- If you persist in trying to answer every difficult question *immediately*, you may find yourself rushing or unable to finish the remainder of the examination.

• When you have finished the examination, return to those unanswered questions.
• Good advice: If at all possible, reread the *entire* examination—and your responses—at least one more time. (This will help you eliminate those errors that result from misreading questions.)

TIP 5

Use your reasoning skills.

SUGGESTIONS
• Answer *all* questions.
• Relate (connect) the question to anything that you studied, wrote in your notebook, or heard your teacher say in class.
• Relate (connect) the question to any film, demonstration, or experiment you saw in class, any project you did, or to anything you may have learned from newspapers, magazines, or television.
• Look over the entire test to see whether one part of it can help you answer another part.

TIP 6

Use your reference tables.

SUGGESTIONS
• You should be familiar with the *content* of each table.
• Frequently, the answers to questions can be found from information contained within these tables.
• Some questions refer to specific tables.
• Other questions do not refer to a table, but can be answered by choosing and using the correct table.
• Be especially familiar with the Periodic Table of the Elements, as well as Reference Table *T*, Equations for Chemistry.

TIP 7

Don't be afraid to guess.

SUGGESTIONS
- Eliminate obvious incorrect choices.
- If still unsure of an answer, make an educated guess.
- There is no penalty for guessing; therefore, answer ALL questions. An omitted answer gets no credit.

TIP 8

Sign the declaration.

SUGGESTIONS
- Be certain that you sign the declaration found on your answer sheet.
- Unless this declaration is signed, your paper cannot be scored.

SUMMARY OF TIPS

1. Be confident and prepared.
2. Read test instructions carefully.
3. Read each question carefully and read each choice before you record your answer.
4. Budget your test time (3 hours).
5. Use your reasoning skills.
6. Use your reference tables.
7. Don't be afraid to guess.
8. Sign the declaration.

HOW TO ANSWER PART *C* (EXTENDED-RESPONSE) QUESTIONS

An *extended-response question* is an examination question that requires the test taker to do more than choose among several responses or fill in a blank. You may need to perform numerical calculations, draw and interpret graphs, and provide extended written responses to a question or a problem.

Part *C* of the New York State Regents Examination in Chemistry contains extended-response questions. This section is designed to provide you with a number of general guidelines for answering them.

Solving Problems Involving Numerical Calculations

To receive full credit you must:
- Provide the appropriate equation(s).
- Substitute values and units into the equation(s).
- Display the answer, with appropriate units and to the correct number of significant figures.
- If the answer is a vector quantity, include its direction.

You should write as legibly as possible. Teachers are human, and nothing irks them more than trying to decipher a careless, messy scrawl. It is also a good idea to identify your answer clearly, either by placing it in a box or by writing the word "answer" next to it.

A final word: If you provide the correct answer but do not show any work, you will not receive any credit for the problem!

The following is a sample problem and its model solution.

PROBLEM

A 5.00-gram object has a density of 4.00 grams per cubic centimeter. Calculate the volume of this object.

SOLUTION

$$d = \frac{m}{V}$$

Rearranging the equation gives

$$V = \frac{m}{d} = \frac{5.00 \text{ g}}{4.00 \text{ g/cm}^3}$$

$$\boxed{V = 1.25 \text{ cm}^3}$$

Graphing Experimental Data

To receive full credit you must:
- Label both axes with the appropriate variables and units.
- Divide the axes so that the data ranges fill the graph as nearly as possible.
- Plot all data points accurately.
- Draw a best-fit line carefully with a straightedge. The line should pass through the origin *only if the data warrant it.*
- If a part of the question requires that the slope be calculated, calculate the slope *from the line,* not from individual data points.

A graph should have a title, and the *independent variable* is usually drawn along the x-axis.

The following is a sample problem and its model solution.

PROBLEM

A student attempts to estimate absolute zero in the following way: He subjects a sample of gas (at constant pressure and mass) to varying temperatures and measures the gas volume at each of the temperatures. The accompanying table contains his experimental data.

Temperature/ °C	Volume/mL
−100	128
−60	148
−40	156
0	204
40	222
80	272
160	310

(a) Using axes that are appropriately labeled and scaled, draw a graph that accurately displays the student's data.
(b) Estimate the student's value for absolute zero by extending the graph to the Celsius temperature at which the volume of the gas is 0 milliliter.

SOLUTION

(a) The first step is to construct an appropriate set of axes if one is not provided. We will assume that you must start from scratch. Since temperature is the independent variable, you need to place it along the x-axis. Also, since absolute zero is −273.15°C, you must scale the axes properly. Here is one appropriate set of axes:

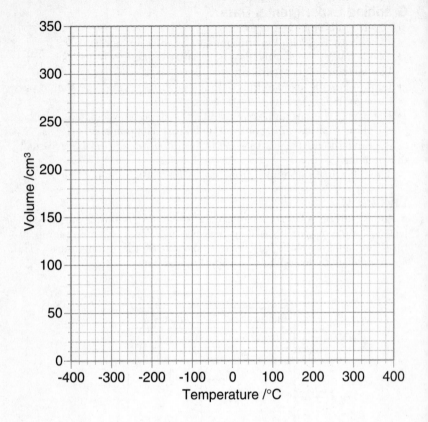

Second, you must plot the data points carefully on the axes as shown below:

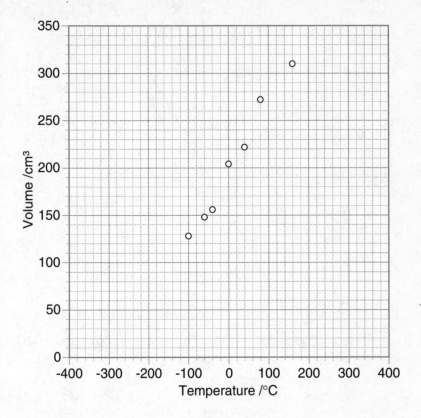

Your third task is to draw the graph. You might be tempted to "connect the dots," but then you would miss the significant relationship between volume and temperature. If you examine *all* of the plotted points, you will note that they fall approximately on a *straight line.*

Therefore, the next step is to draw a *best-fit* straight-line graph. This is a graph in which the data points are most closely distributed on both sides of the line. The accompanying graph shows the best-fit straight line.

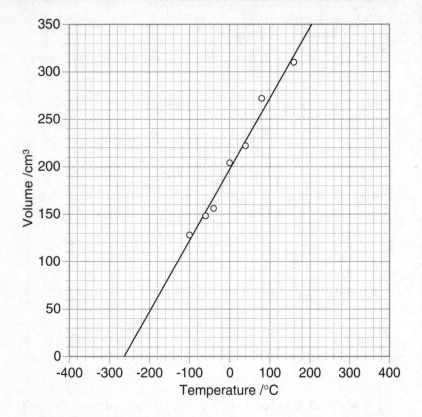

Note that the graph extending beyond the data points (above and below) is a *dashed* line. Such extensions are known as *extrapolated data*, and they are based on the assumption that the gas will continue to behave as it did within the experimental range for which the student collected data.

(You may ask: Why *this* particular line? It seems as though many lines could have been drawn using the plotted data. Actually, there is only *one* best-fit straight line, and it is calculated by using an advanced statistical technique known as *linear regression*. This technique was used to draw the line shown above. At this point, it is sufficient for you to provide an "eyeball" estimate of the best-fit straight line.)

(b) Your final task is to inspect the graph closely and to estimate absolute zero. The calculated value is –263°C. (This corresponds to an experimental error of 3.7%).

Drawing Diagrams

To receive full credit you must:

• Draw your diagrams neatly and label them clearly.
• Bring a straightedge and a protractor with you so that you can draw neat, accurate diagrams.

The accompanying diagram represents a zinc–copper electrochemical cell containing an agar–KCl salt bridge. This is the type of diagram you may be asked to draw as part of an examination question.

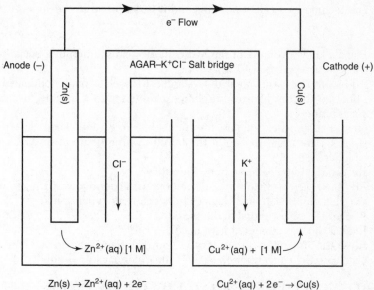

$$Zn(s) \rightarrow Zn^{2+}(aq) + 2e^-$$
(oxidation half-reaction)

$$Cu^{2+}(aq) + 2e^- \rightarrow Cu(s)$$
(reduction half-reaction)

Net reaction: $Zn(s) + Cu^{2+}(aq) \rightarrow Cu(s) + Zn^{2+}(aq)$

Writing a Free-Response Answer

To receive full credit you must:

• Use complete, clear sentences that make sense to the reader.
• Use correct chemistry in your explanations.

A sample question and acceptable and unacceptable answers are given below.

QUESTION

Describe in detail the technique used to determine the concentration of a dilute hydrochloric acid solution using a dilute sodium hydroxide solution of known concentration.

Name any equipment or other chemicals that you would use. *You need not describe any mathematical calculations.*

MODEL ANSWER

Since this question requires an extended answer containing a series of steps, we decide to use an outline form.

The technique is called *titration*, and it is described in the steps given below.

- Place a known volume of the hydrochloric acid solution in a beaker of suitable size.
- Add a drop or two of an acid–base indicator such as phenolphthalein.
- Pour the sodium hydroxide solution carefully into a burette, using a small funnel.
- Open the stopcock to allow the trapped air to escape. Then close the stopcock and wipe the tip of the burette with a tissue to remove any clinging liquid.
- Record the initial volume in the burette.
- Add the base slowly to the acid solution, with continuous stirring, until the phenolphthalein just changes from colorless to faint pink.
- Record the final volume in the burette.
- Repeat the experiment at least once.
- Rinse the apparatus with water to remove all traces of acid and base.
- Calculate the concentration of the hydrochloric acid solution.

WRITING AND BALANCING EQUATIONS IN A CHEMISTRY EXAMINATION

For all of the suggestions that follow, the reaction that occurs between aqueous solutions of sodium sulfate and barium nitrate is used as an example and is referred to as the given reaction.

- If you are asked to write a *word equation*, be certain to include the correct names of the reactants and products and their phases in the reaction.

For the given reaction, the word equation is

barium nitrate(aq) + sodium sulfate(aq) →
barium sulfate(s) + sodium nitrate(aq)

- If you are asked to write a *balanced* equation, you are usually expected to balance using *smallest whole-number coefficients*.

For the given reaction, the balanced equation is

$$Ba(NO_3)_2(aq) + Na_2SO_4(aq) \rightarrow BaSO_4(s) + 2NaNO_3(aq)$$

- If you are asked to write an *ionic* equation occurring in aqueous solution, you must reduce everything to its component ions except *insoluble* compounds, such as $BaSO_4(s)$, and (of course!) *covalently* bonded substances, such as $H_2O(\ell)$.

For the given reaction, the ionic equation is

$$Ba^{2+}(aq) + 2NO_3^-(aq) + 2Na^+(aq) + SO_4^{2-}(aq) \rightarrow$$
$$BaSO_4(s) + 2Na^+(aq) + 2NO_3^-(aq)$$

- If you are asked to write a *net* ionic equation, you should omit all *spectator ions*, that is, all ions appearing *unchanged on both sides of the equation*.

For the given reaction, the net ionic equation is

$$Ba^{2+}(aq) + SO_4^{2-}(aq) \rightarrow BaSO_4(s)$$

What to Expect on the Chemistry Examination

FORMAT OF THE CHEMISTRY EXAMINATION

The chemistry examination will be 3 hours long and will include three parts: A, B, and C. You should be prepared to answer questions in multiple-choice format, as well as answer questions that require a more extended response.

Questions will be content- and skills-based, and you may be required to graph data, complete a data table, label or draw diagrams, design experiments, make calculations, or write short or extended responses.

In addition, you may be required to hypothesize, interpret, analyze, evaluate data, or apply your scientific knowledge and skills to real-world situations.

Some of the questions will require use of the 2002 edition of the Reference Tables for Physical Setting/Chemistry.

[In the future, a Part D will be added, which will focus on assessment of laboratory skills. As more information becomes available, the New York State Education Department will inform schools of the development status of the performance test.]

You will be required to answer ALL of the questions on the Physical Setting/Chemistry Regents examination.

Physical Setting/Chemistry Regents Examination Format

PART	ITEM TYPE(S)	DESCRIPTION OF THE ITEMS	APPROXIMATE PERCENT OF TOTAL TEST RAW SCORE
A	Multiple-choice questions	Content-based questions assessing your knowledge and understanding of core material	35
B	Multiple-choice and constructed-response questions	Content- and skills-based questions assessing your ability to apply, analyze, and evaluate material	30
C	Constructed-response and/or extended constructed-response questions	Content-based and application questions assessing your ability to apply knowledge of science concepts and skills to address real-world situations	20

The maximum *raw* score on the examination is 85 points. A teacher's chart will be provided for converting your *raw* score to a *scaled* score that has a maximum of 100 points. A sample conversion table taken from the June 2004 Regents Chemistry Examination is shown below:

Sample Conversion Table

RAW SCORE	SCALED SCORE	RAW SCORE	SCALED SCORE	RAW SCORE	SCALED SCORE	RAW SCORE	SCALED SCORE
85	100	63	74	41	58	19	35
84	98	62	73	40	57	18	34
83	97	61	73	39	56	17	33
82	95	60	72	38	55	16	31
81	94	59	71	37	55	15	30
80	93	58	70	36	54	14	28
79	91	57	69	35	53	13	26
78	90	56	69	34	52	12	25
77	89	55	68	33	51	11	23
76	87	54	67	32	50	10	21
75	86	53	66	31	49	9	19
74	85	52	66	30	48	8	17
73	84	51	65	29	47	7	15
72	83	50	64	28	46	6	13
71	82	49	64	27	45	5	11
70	81	48	63	26	44	4	9
69	80	47	62	25	43	3	7
68	79	46	61	24	42	2	5
67	78	45	61	23	41	1	2
66	77	44	60	22	39	0	0
65	76	43	59	21	38		
64	75	42	58	20	37		

The table is used to convert the number of points you actually received on the examination (your "raw" score) to your final score on the examination (your "scaled" score). **Note that this table will change from one examination to another.**

TOPICS COVERED ON THE CHEMISTRY EXAMINATION

All of the questions on the Chemistry examination will test major understandings, skills, and real-world applications drawn from the following 11 subject areas:

M.   Math Skills
I.   Atomic Concepts
II.   Periodic Table
III.   Moles/Stoichiometry
IV.   Chemical Bonding
V.   Physical Behavior of Matter
VI.   Kinetics/Equilibrium
VII.   Organic Chemistry
VIII. Oxidation–Reduction
IX.   Acids, Bases, and Salts
X.   Nuclear Chemistry

It is suggested that you read the *Topic Outline* found on pages 24–43 in order to learn the exact nature of the material that is subject to testing.

New York State Physical Setting/ Chemistry Core

TOPIC OUTLINE

The Topic Outline on pages 24-43 is taken from Appendix B of the New York State Physical Setting/Chemistry Core. All Regents Chemistry Examinations are based on this core. The topic outline is divided into 11 sections:

M.	Math Skills
I.	Atomic Concepts
II.	Periodic Table
III.	Moles/Stoichiometry
IV.	Chemical Bonding
V.	Physical Behavior of Matter
VI.	Kinetics/Equilibrium
VII.	Organic Chemistry
VIII.	Oxidation–Reduction
IX.	Acids, Bases, and Salts
X.	Nuclear Chemistry

Each section contains one or more of the following items:

- The *Major Understandings* that you must have mastered for the examination
- The *Skills* that you need to be able to demonstrate during the examination
- The *Real-World Connections* that relate chemistry concepts to the world around you.

M. Mathematics Skills Needed for Chemistry

M.1 Organize, graph, and analyze data gathered from laboratory activities or other sources.
- Identify independent and dependent variables.
- Create appropriate axes with label and scale.
- Identify graph points clearly.

M.2 Interpret a graph constructed from experimentally determined data.
- Identify direct and inverse relationships.
- Apply data showing trends to predict information.

M.3 Measure and record experimental data and use the data in calculations.
- Choose appropriate measurement scales and use units in recording.
- Show mathematical work stating formula and steps for solution.
- Estimate answers.
- Use appropriate equations and significant digits.
- Identify relationships within variables from data tables.
- Calculate percent error.

M.4 Recognize and convert various scales of measurement.
- Convert between Celsius (°C) and Kelvin (K).
- Convert among kilometers (km), meters (m), centimeters (cm), and millimeters (mm).
- Convert between grams (g) and kilograms (kg).
- Convert between kilopascals (kPa) and atmospheres (atm).

I. Atomic Concepts

	MAJOR UNDERSTANDINGS	SKILLS The student should be able to:	REAL-WORLD CONNECTIONS
I.1	The modern model of the atom has evolved over a long period of time through the work of many scientists.	relate experimental evidence to models of the atom	
I.2	Each atom has a nucleus, with an overall positive charge, surrounded by negatively charged electrons.	use models to describe the structure of an atom	
I.3	Subatomic particles contained in the nucleus include protons and neutrons.		
I.4	The proton is positively charged, and the neutron has no charge. The electron is negatively charged.		
I.5	Protons and electrons have equal but opposite charges. The number of protons is equal to the number of electrons in an atom.	determine the number of protons or electrons in an atom or ion when given one of these values	
I.6	The mass of each proton and each neutron is approximately equal to one atomic mass unit. An electron is much less massive than a proton or neutron.	calculate the mass of an atom, the number of neutrons, or the number of protons, given the other two values	♦ lasers
I.7	In the wave-mechanical model (electron cloud), the electrons are in orbitals, which are defined as regions of most probable electron location (ground state).		
I.8	Each electron in an atom has its own distinct amount of energy.		
I.9	When an electron in an atom gains a specific amount of energy, the electron is at a higher energy state (excited state).	distinguish between ground state and excited state electron configurations, e.g., 2–8–2 vs. 2–7–3	

I. Atomic Concepts

	MAJOR UNDERSTANDINGS	SKILLS The student should be able to:	REAL-WORLD CONNECTIONS
I.10	When an electron returns from a higher energy state to a lower energy state, a specific amount of energy is emitted. This emitted energy can be used to identify an element.	identify an element by comparing its bright-line spectrum to given spectra	♦ flame tests ♦ neon lights ♦ fireworks ♦ forensic analysis ♦ spectral analysis of stars
I.11	The outermost electrons in an atom are called the valence electrons. In general, the number of valence electrons affects the chemical properties of an element.	draw a Lewis electron-dot structure of an atom distinguish between valence and non-valence electrons, given an electron configuration, e.g., 2–8–2	
I.12	Atoms of an element that contain the same number of protons but a different number of neutrons are called isotopes of that element.		
I.13	The average atomic mass of an element is the weighted average of the masses of its naturally occurring isotopes.	given an atomic mass, determine the most abundant isotope calculate the atomic mass of an element, given the masses and ratios of naturally occurring isotopes	

II. Periodic Table

II.1	The placement or location of an element on the Periodic Table gives an indication of physical and chemical properties of that element. The elements on the Periodic Table are arranged in order of increasing atomic number.	explain the placement of an unknown element in the Periodic Table based on its properties	♦ similar properties and uses for elements in the same family ♦ characteristics of a class of elements are similar

II. Periodic Table

	MAJOR UNDERSTANDINGS	SKILLS The student should be able to:	REAL-WORLD CONNECTIONS
II.2	The number of protons in an atom (atomic number) identifies the element. The sum of the protons and neutrons in an atom (mass number) identifies an isotope. Common notations that represent isotopes include: ^{14}C, $^{14}_{6}C$, carbon-14, C-14.	interpret and write isotopic notation	
II.3	Elements can be classified by their properties and located on the Periodic Table as metals, nonmetals, metalloids (B, Si, Ge, As, Sb, Te), and noble gases.	classify elements as metals, nonmetals, metalloids, or noble gases by their properties	♦ similar properties and uses for elements in the same family
II.4	Elements can be differentiated by their physical properties. Physical properties of substances, such as density, conductivity, malleability, solubility, and hardness, differ among elements.	describe the states of the elements at STP	♦ uses of different elements, e.g., use of semiconductors in solid state electronics and computer technology ♦ alloys as superconductors
II.5	Elements can be differentiated by chemical properties. Chemical properties describe how an element behaves during a chemical reaction.		♦ metallurgy ♦ recovery of metals
II.6	Some elements exist as two or more forms in the same phase. These forms differ in their molecular or crystal structure and hence in their properties.		♦ different properties for each allotrope: ∞ oxygen gas vs. ozone ∞ coal vs. graphite vs. diamond vs. buckminsterfullerene
II.7	For Groups 1, 2, and 13–18 on the Periodic Table, elements within the same group have the same number of valence electrons (helium is an exception) and, therefore, similar chemical properties.	determine the group of an element, given the chemical formula of a compound, e.g., XCl or XCl_2	

II. Periodic Table

	MAJOR UNDERSTANDINGS	SKILLS The student should be able to:	REAL-WORLD CONNECTIONS
II.8	The succession of elements within the same group demonstrates characteristic trends: differences in atomic radius, ionic radius, electronegativity, first ionization energy, and metallic/nonmetallic properties.	compare and contrast properties of elements within a group or a period for Groups 1, 2, 13–18 on the Periodic Table	
II.9	The succession of elements across the same period demonstrates characteristic trends: differences in atomic radius, ionic radius, electronegativity, first ionization energy, and metallic/nonmetallic properties.		

III. Moles/Stoichiometry

	MAJOR UNDERSTANDINGS	SKILLS The student should be able to:	REAL-WORLD CONNECTIONS
III.1	A compound is a substance composed of two or more different elements that are chemically combined in a fixed proportion. A chemical compound can be broken down by chemical means. A chemical compound can be represented by a specific chemical formula and assigned a name based on the IUPAC system.		♦ reading food and beverage labels (consumer chemistry)
III.2	Types of chemical formulas include empirical, molecular, and structural.		

III. Moles/Stoichiometry

	MAJOR UNDERSTANDINGS	SKILLS The student should be able to:	REAL-WORLD CONNECTIONS
III.3	The empirical formula of a compound is the simplest whole-number ratio of atoms of the elements in a compound. It may be different from the molecular formula, which is the actual ratio of atoms in a molecule of that compound.	determine the molecular formula, given the empirical formula and molecular mass determine the empirical formula from a molecular formula	
III.4	In all chemical reactions there is a conservation of mass, energy, and charge.	interpret balanced chemical equations in terms of conservation of matter and energy	
III.5	A balanced chemical equation represents conservation of atoms. The coefficients in a balanced chemical equation can be used to determine mole ratios in the reaction.	balance equations, given the formulas for reactants and products interpret balanced chemical equations in terms of conservation of matter and energy create and use models of particles to demonstrate balanced equations calculate simple mole–mole stoichiometry problems, given a balanced equation	
III.6	The formula mass of a substance is the sum of the atomic masses of its atoms. The molar mass (gram-formula mass) of a substance equals one mole of that substance.	calculate the formula mass and the gram-formula mass	
III.7	The percent composition by mass of each element in a compound can be calculated mathematically.	determine the number of moles of a substance, given its mass determine the mass of a given number of moles of a substance	

III. Moles/Stoichiometry

	MAJOR UNDERSTANDINGS	SKILLS The student should be able to:	REAL-WORLD CONNECTIONS
III.8	Types of chemical reactions include synthesis, decomposition, single replacement, and double replacement.	identify types of chemical reactions	◆ recovery of metals from ores ◆ electroplating ◆ corrosion ◆ precipitation reactions ◆ dangers of mixing household chemicals together, e.g., bleach and ammonia ◆ electrolysis of active metal compounds ◆ explosives (inflation of air bags)

IV. Chemical Bonding

IV.1	Compounds can be differentiated by their chemical and physical properties.	distinguish among ionic, molecular, and metallic substances, given their properties	
IV.2	Two major categories of compounds are ionic and molecular (covalent) compounds.		
IV.3	Chemical bonds are formed when valence electrons are: transferred from one atom to another (ionic); shared between atoms (covalent); mobile within a metal (metallic).	demonstrate bonding concepts using Lewis dot structures representing valence electrons: transferred (ionic bonding); shared (covalent bonding); in a stable octet	◆ photosynthesis ◆ DNA bonding
IV.4	In a multiple covalent bond, more than one pair of electrons are shared between two atoms. Unsaturated organic compounds contain at least one double or triple bond.		
IV.5	Molecular polarity can be determined by the shape and distribution of the charge. Symmetrical (nonpolar) molecules include CO_2, CH_4, and diatomic elements. Asymmetrical (polar) molecules include HCl, NH_3, H_2O.		

IV. Chemical Bonding

	MAJOR UNDERSTANDINGS	SKILLS The student should be able to:	REAL-WORLD CONNECTIONS
IV.6	When an atom gains one or more electrons, it becomes a negative ion and its radius increases. When an atom loses one or more electrons, it becomes a positive ion and its radius decreases.		♦ saturated vs. unsaturated compounds—health connections
IV.7	When a bond is broken, energy is absorbed. When a bond is formed, energy is released.		
IV.8	Atoms attain a stable valence electron configuration by bonding with other atoms. Noble gases have stable valence electron configurations and tend not to bond.	determine the noble gas configuration an atom will achieve when bonding	
IV.9	Physical properties of substances can be explained in terms of chemical bonds and intermolecular forces. These properties include conductivity, malleability, solubility, hardness, melting point, and boiling point.		
IV.10	Electron-dot diagrams (Lewis structures) can represent the valence electron arrangement in elements, compounds, and ions.	demonstrate bonding concepts, using Lewis dot structures representing valence electrons: transferred (ionic bonding); shared (covalent bonding); in a stable octet	♦ free radicals
IV.11	Electronegativity indicates how strongly an atom of an element attracts electrons in a chemical bond. Electronegativity values are assigned according to arbitrary scales.		
IV.12	The electronegativity difference between two bonded atoms is used to assess the degree of polarity in the bond.	distinguish between nonpolar covalent bonds (two of the same nonmetals) and polar covalent bonds	

IV. Chemical Bonding

	MAJOR UNDERSTANDINGS	SKILLS The student should be able to:	REAL-WORLD CONNECTIONS
IV.13	Metals tend to react with nonmetals to form ionic compounds. Nonmetals tend to react with other nonmetals to form molecular (covalent) compounds. Ionic compounds containing polyatomic ions have both ionic and covalent bonding.		

V. Physical Behavior of Matter

V.1	Matter is classified as a pure substance or as a mixture of substances.		

V.2	The three phases of matter (solids, liquids, and gases) have different properties.	use a simple particle model to differentiate properties of a solid, a liquid, and a gas	◆ common everyday examples of solids, liquids, and gases
			◆ nature of H_2O in our environment
			◆ solids
			  ∞ metallic
			  ∞ crystalline
			  ∞ amorphous (quartz glass, opals)
			  ∞ solid state
			◆ liquids
			  ∞ surface tension
			  ∞ capillary
			  ∞ viscosity
			◆ gases
			  ∞ real and ideal gases

V.3	A pure substance (element or compound) has a constant composition and has constant properties throughout a given sample and from sample to sample.	use particle models/diagrams to differentiate elements, compounds, and mixtures	
V.4	Elements are substances that are composed of atoms that have the same atomic number. Elements cannot be broken down by chemical change.		

V. Physical Behavior of Matter

	MAJOR UNDERSTANDINGS	SKILLS The student should be able to:	REAL-WORLD CONNECTIONS
V.5	Mixtures are composed of two or more different substances that can be separated by physical means. When different substances are mixed together, a homogeneous or heterogeneous mixture is formed.		♦ alloys ♦ separation by filtration, distillation, desalination, crystallization, extraction, chromatography ♦ water quality testing ♦ colloids ♦ emulsifiers (making ice cream) ♦ sewage treatment
V.6	The proportions of components in a mixture can be varied. Each component in a mixture retains its original properties.		
V.7	Differences in properties such as density, particle size, molecular polarity, boiling point and freezing point, and solubility permit physical separation of the components of the mixture.	describe the process and use of filtration, distillation, and chromatography in the separation of a mixture	
V.8	A solution is a homogeneous mixture of a solute dissolved in a solvent. The solubility of a solute in a given amount of solvent is dependent on the temperature, the pressure, and the chemical natures of the solute and solvent.	interpret and construct solubility curves use solubility curves to distinguish saturated, supersaturated, and unsaturated solutions apply the adage "like dissolves like" to real-world situations	♦ degrees of saturation of solutions ♦ dry cleaning
V.9	The concentration of a solution may be expressed as: molarity (M), percent by volume, percent by mass, or parts per million (ppm).	describe the preparation of a solution, given the molarity interpret solution concentration data calculate solution concentrations in molarity (M), percent mass, and parts per million (ppm)	

V. Physical Behavior of Matter

	MAJOR UNDERSTANDINGS	SKILLS The student should be able to:	REAL-WORLD CONNECTIONS
V.10	The addition of a nonvolatile solute to a solvent causes the boiling point of the solvent to increase and the freezing point of the solvent to decrease. The greater the concentration of solute particles, the greater the effect.		♦ salting an icy sidewalk ♦ ice cream making ♦ antifreeze/engine coolant ♦ airplane deicing ♦ cooking pasta
V.11	Energy can exist in different forms, such as chemical, electrical, electromagnetic, heat, mechanical, and nuclear.		
V.12	Heat is a transfer of energy (usually thermal energy) from a body of higher temperature to a body of lower temperature. Thermal energy is associated with the random motion of atoms and molecules.	distinguish between heat energy and temperature in terms of molecular motion and amount of matter qualitatively interpret heating and cooling curves in terms of changes in kinetic and potential energy, heat of vaporization, heat of fusion, and phase changes	
V.13	Temperature is a measure of the average kinetic energy of the particles in a sample of matter. Temperature is not a form of energy.	distinguish between heat energy and temperature in terms of molecular motion and amount of matter explain phase changes in terms of the changes in energy and intermolecular distance	
V.14	The concept of an ideal gas is a model to explain behavior of gases. A real gas is most like an ideal gas when the real gas is at low pressure and high temperature.		♦ Earth's primitive atmosphere ♦ use of models to explain something that cannot be seen

V. Physical Behavior of Matter

		MAJOR UNDERSTANDINGS	SKILLS The student should be able to:	REAL-WORLD CONNECTIONS
V.15		Kinetic molecular theory (KMT) for an ideal gas states all gas particles: ♦ are in random, constant, straight-line motion ♦ are separated by great distances relative to their size; the volume of gas particles is considered negligible ♦ have no attractive forces between them ♦ have collisions that may result in a transfer of energy between particles, but the total energy of the system remains constant.		
V.16		Collision theory states that a reaction is most likely to occur if reactant particles collide with the proper energy and orientation.		
V.17		Kinetic molecular theory describes the relationships of pressure, volume, temperature, velocity, and frequency and force of collisions among gas molecules.	explain the gas laws in terms of KMT solve problems, using the combined gas law	♦ structure and composition of Earth's atmosphere (variations in pressure and temperature)
V.18		Equal volumes of gases at the same temperature and pressure contain an equal number of particles.	convert temperatures in Celsius degrees (°C) to kelvins (K), and kelvins to Celsius degrees	
V.19		The concepts of kinetic and potential energy can be used to explain physical processes that include: fusion (melting); solidification (freezing); vaporization (boiling, evaporation), condensation, sublimation, and deposition.	qualitatively interpret heating and cooling curves in terms of changes in kinetic and potential energy, heat of vaporization, heat of fusion, and phase changes calculate the heat involved in a phase or temperature change for a given sample of matter explain phase change in terms of the changes in energy and intermolecular distances	♦ weather processes ♦ greenhouse gases

V. Physical Behavior of Matter

	MAJOR UNDERSTANDINGS	SKILLS The student should be able to:	REAL-WORLD CONNECTIONS
V.20	A physical change results in the rearrangement of existing particles in a substance. A chemical change results in the formation of different substances with changed properties.		
V.21	Chemical and physical changes can be exothermic or endothermic.	distinguish between endothermic and exothermic reactions, using energy terms in a reaction equation, ΔH, potential energy diagrams or experimental data	♦ calorimetry
V.22	The structure and arrangement of particles and their interactions determine the physical state of a substance at a given temperature and pressure.	use a simple particle model to differentiate properties of solids, liquids, and gases	
V.23	Intermolecular forces created by the unequal distribution of change result in varying degrees of attraction between molecules. Hydrogen bonding is an example of a strong intermolecular force.	explain vapor pressure, evaporation rate, and phase changes in terms of intermolecular forces	♦ refrigeration ♦ meniscus (concave/convex) ♦ capillary action ♦ surface tension
V.24	Physical properties of substances can be explained in terms of chemical bonds and intermolecular forces. These properties include conductivity, malleability, solubility, hardness, melting point, and boiling point.	compare the physical properties of substances based upon chemical bonds and intermolecular forces	

VI. Kinetics/Equilibrium

VI.1	Collision theory states that a reaction is most likely to occur if reactant particles collide with the proper energy and orientation.	use collision theory to explain how various factors, such as temperature, surface area, and concentration, influence the rate of reaction	♦ synthesis of compounds

VI. Kinetics/Equilibrium

	MAJOR UNDERSTANDINGS	SKILLS The student should be able to:	REAL-WORLD CONNECTIONS
VI.2	The rate of a chemical reaction depends on several factors: temperature, concentration, nature of reactants, surface area, and the presence of a catalyst.		♦ catalysts and inhibitors
VI.3	Some chemical and physical changes can reach equilibrium.	identify examples of physical equilibria as solution equilibrium and phase equilibrium, including the concept that a saturated solution is at equilibrium	♦ balloons
VI.4	At equilibrium, the rate of the forward reaction equals the rate of the reverse reaction. The measurable quantities of reactants and products remain constant at equilibrium.	describe the concentration of particles and rates of opposing reactions in an equilibrium system	
VI.5	LeChâtelier's principle can be used to predict the effect of stress (change in pressure, volume, concentration, and temperature) on a system at equilibrium.	qualitatively describe the effect of stress on equilibrium, using LeChâtelier's principle	♦ Haber process
VI.6	Energy released or absorbed by a chemical reaction can be represented by a potential energy diagram.	read and interpret potential energy diagrams: PE of reactants and products, activation energy (with or without a catalyst), heat of reaction	
VI.7	Energy released or absorbed by a chemical reaction (heat of reaction) is equal to the difference between the potential energy of the products and the potential energy of the reactants.		♦ burning fossil fuels ♦ photosynthesis ♦ production of photochemical smog
VI.8	A catalyst provides an alternate reaction pathway that has a lower activation energy than an uncatalyzed reaction.		♦ enzymes in the human body

VI. Kinetics/Equilibrium

	MAJOR UNDERSTANDINGS	SKILLS The student should be able to:	REAL-WORLD CONNECTIONS
VI.9	Entropy is a measure of the randomness or disorder of a system. A system with greater disorder has greater entropy.	compare the entropy of phases of matter	♦ relationship to phase change
VI.10	Systems in nature tend to undergo changes toward lower energy and higher entropy.		♦ chaos therapy—randomness vs. order

VII. Organic Chemistry

	MAJOR UNDERSTANDINGS	SKILLS The student should be able to:	REAL-WORLD CONNECTIONS
VII.1	Organic compounds contain carbon atoms that bond to one another in chains, rings, and networks to form a variety of structures. Organic compounds can be named using the IUPAC system.	classify an organic compound based on its structural or condensed structural formula	♦ biochemical molecules-formation of carbohydrates, proteins, starches, fats, and nucleic acids ♦ synthetic polymers-polyethylene (plastic bags, toys), polystyrene (cups, insulation), polypropylene (carpets, bottles), polytetrafluoroethylene (nonstick surfaces—TEFLON), and polyacrilonitrile (yarns, fabrics, wigs) ♦ disposal problems of synthetic polymers
VII.2	Hydrocarbons are compounds that contain only carbon and hydrogen. Saturated hydrocarbons contain only single carbon–carbon bonds. Unsaturated hydrocarbons contain at least one multiple carbon–carbon bond.	draw structural formulas for alkanes, alkenes, and alkynes containing a maximum of ten carbon atoms	
VII.3	Organic acids, alcohols, esters, aldehydes, ketones, ethers, halides, amines, amides, and amino acids are types of organic compounds that differ in their structures. Functional groups impart distinctive physical and chemical properties to organic compounds.	classify an organic compound based on its structural or condensed structural formula draw a structural formula with the functional group(s) on a straight chain hydrocarbon backbone, when given the correct IUPAC name for the compound	♦ making perfume ♦ wine production ♦ nuclear magnetic resonance spectroscopy (NMR), (MRI) ♦ dyes ♦ cosmetics ♦ odors (esters)

VII. Organic Chemistry

	MAJOR UNDERSTANDINGS	SKILLS The student should be able to:	REAL-WORLD CONNECTIONS
VII.4	Isomers of organic compounds have the same molecular formula, but different structures and properties.		♦ types, varieties, uses of organic compounds ♦ organic isomers
VII.5	In a multiple covalent bond, more than one pair of electrons are shared between two atoms. Unsaturated organic compounds contain at least one double or triple bond.		♦ saturated vs. unsaturated compounds—health connections
VII.6	Types of organic reactions include: addition, substitution, polymerization, esterification, fermentation, saponification, and combustion.	identify types of organic reactions determine a missing reactant or product in a balanced equation	♦ saponification—making soap ♦ polymerization—formation of starches ♦ fermentation—alcohol production ♦ combustion of fossil fuels ♦ cellular respiration

VIII. Oxidation–Reduction

	MAJOR UNDERSTANDINGS	SKILLS The student should be able to:	REAL-WORLD CONNECTIONS
VIII.1	An oxidation–reduction (redox) reaction involves transfer of electrons (e^-).	determine a missing reactant or product in a balanced equation	♦ electrochemical cells ♦ corrosion ♦ electrolysis ♦ photography ♦ rusting
VIII.2	Reduction is the gain of electrons.		♦ smelting ♦ leaching (refining of gold) ♦ thermite reactions (reduction of metal oxides, e.g., aluminum)
VIII.3	A half-reaction can be written to represent reduction.	write and balance half-reactions for oxidation and reduction of free elements and their monatomic ions	
VIII.4	Oxidation is the loss of electrons.		♦ recovery of active non-metals (I_2)

VIII. Oxidation–Reduction

	MAJOR UNDERSTANDINGS	SKILLS The student should be able to:	REAL-WORLD CONNECTIONS
VIII.5	A half-reaction can be written to represent oxidation.		
VIII.6	In a redox reaction, the number of electrons lost is equal to the number of electrons gained.		
VIII.7	Oxidation numbers (states) can be assigned to atoms and ions. Changes in oxidation numbers indicate that oxidation and reduction have occurred.		
VIII.8	An electrochemical cell can be either voltaic or electrolytic. In an electrochemical cell, oxidation occurs at the anode and reduction at the cathode.	compare and contrast voltaic and electrolytic cells	♦ patina (copper—Statue of Liberty)
VIII.9	A voltaic cell spontaneously converts chemical energy to electrical energy.	identify and label the parts of a voltaic cell (cathode, anode, salt bridge) and direction of electron flow, given the reaction equation use a table of reduction potentials to determine whether a redox reaction is spontaneous	
VIII.10	An electrolytic cell requires electrical energy to produce chemical change. This process is known as electrolysis.	identify and label the parts of an electrolytic cell (anode, cathode) and direction of electron flow, given the reaction equation	♦ metallurgy of iron and steel ♦ electroplating

IX. Acids, Bases, and Salts

IX.1	Behavior of many acids and bases can be explained by the Arrhenius theory. Arrhenius acids and bases are electrolytes.	given properties, identify substances as Arrhenius acids or Arrhenius bases	

IX. Acids, Bases, and Salts

	MAJOR UNDERSTANDINGS	SKILLS The student should be able to:	REAL-WORLD CONNECTIONS
IX.2	An electrolyte is a substance that, when dissolved in water, forms a solution capable of conducting an electric current. The ability of a solution to conduct an electric current depends on the concentration of ions.		
IX.3	Arrhenius acids yield H^+ (hydrogen ion) as the only positive ion in aqueous solution. The hydrogen ion may also be written as H_3O^+, hydronium ion.		
IX.4	Arrhenius bases yield OH^- (hydroxide ion) as the only negative ion in an aqueous solution.		♦ cleaning agents
IX.5	In the process of neutralization, an Arrhenius acid and an Arrhenius base react to form salt and water.	write simple neutralization reactions when given the reactants	
IX.6	Titration is a laboratory process in which a volume of solution of known concentration is used to determine the concentration of another solution.	calculate the concentration or volume of a solution, using titration data	
IX.7	There are alternate acid–base theories. One such theory states that an acid is an H+ donor and a base is an H+ acceptor.		
IX.8	The acidity and alkalinity of an aqueous solution can be measured by its pH value. The relative level of acidity or alkalinity of a solution can be shown by using indicators.	interpret changes in acid–base indicator color identify solutions as acid, base, or neutral based upon the pH	♦ acid rain ♦ household chemicals ♦ buffers ♦ swimming pool chemistry ♦ blood acidosis/alkalosis

IX. Acids, Bases, and Salts

	MAJOR UNDERSTANDINGS	SKILLS The student should be able to:	REAL-WORLD CONNECTIONS
IX.9	On the pH scale, each decrease of one unit of pH represents a tenfold increase in hydronium ion concentration.		

X. Nuclear Chemistry

	MAJOR UNDERSTANDINGS	SKILLS The student should be able to:	REAL-WORLD CONNECTIONS
X.1	Stability of isotopes is based on the ratio of the neutrons and protons in its nucleus. Although most nuclei are stable, some are unstable and spontaneously decay emitting radiation.		
X.2	Each radioactive isotope has a specific mode and rate of decay (half-life).	calculate the initial amount, the fraction remaining, or the half-life of a radioactive isotope, given two of the three variables	♦ radioactive dating
X.3	A change in the nucleus of an atom that converts it from one element to another is called transmutation. This can occur naturally or can be induced by the bombardment of the nucleus of high-energy particles.		♦ nuclear fission and fusion reactions that release energy ♦ radioisotopes, tracers, transmutation ♦ man-made elements
X.4	Spontaneous decay can involve the release of alpha particles, beta particles, positrons, and/or gamma radiation from the nucleus of an unstable isotope. These emissions differ in mass, charge, ionizing power, and penetrating power.	determine decay mode and write nuclear equations showing alpha and beta decay	
X.5	Nuclear reactions include natural and artificial transmutation, fission, and fusion.	compare and contrast fission and fusion reactions	
X.6	There are benefits and risks associated with fission and fusion reactions.		

X. Nuclear Chemistry

	MAJOR UNDERSTANDINGS	SKILLS The student should be able to:	REAL-WORLD CONNECTIONS
X.7	Nuclear reactions can be represented by equations that include symbols that represent atomic nuclei (with the mass number and atomic number), subatomic particles (with mass number and charge), and/or emissions such as gamma radiation.	complete nuclear equations; predict missing particles from nuclear equations	
X.8	Energy released in a nuclear reaction (fission or fusion) comes from the fractional amount of mass converted into energy. Nuclear changes convert matter into energy.		◆ production of nuclear power ∞ fission ∞ fusion (breeder reactors) ◆ cost–benefit analysis among various types of power production
X.9	Energy released during nuclear reactions is much greater than the energy released during chemical reactions.		
X.10	There are inherent risks associated with radioactivity and the use of radioactive isotopes. Risks can include biological exposure, long-term storage and disposal, and nuclear accidents.		◆ nuclear waste ◆ radioactive pollution
X.11	Radioactive isotopes have many beneficial uses. Radioactive isotopes are used in medicine and industrial chemistry, e.g., radioactive dating, tracing chemical and biological processes, industrial measurement, nuclear power, and detection and treatment of diseases.	identify specific uses of some common radioisotopes, such as: I-131 in diagnosing and treating thyroid disorders; C-14 to C-12 ratio in dating living organisms; U-238 to Pb-206 ratio in dating geological formations; Co-60 in treating cancer	◆ use of radioactive tracers ◆ radiation therapy ◆ irradiated food

QUESTION INDEX

What follows is an index to the examination questions that are explained in this book. The questions are indexed according to the *Sequence numbers given in the Topic Outline* found on pages 24–43.

Some questions embrace more than one topic; these questions are marked with a dagger (†).

SEQUENCE	AUGUST 2002	JANUARY 2003	JUNE 2003	AUGUST 2003	JUNE 2004	AUGUST 2004	JUNE 2005	AUGUST 2005

III. Moles/Stoichiometry (continued)

SEQUENCE	AUGUST 2002	JANUARY 2003	JUNE 2003	AUGUST 2003	JUNE 2004	AUGUST 2004	JUNE 2005	AUGUST 2005
III.3		†13	8	42	38, 54, 70	37	37	52
III.4	9			10	†55	8		30
III.5	47	42, 48	20, 59	39, 54	7, 52, 53, †55, 66	52, 66	54, 74	48, 68
III.6	56 b				6	61		51, 69
III.7		8, 22	10	8	8		77	31
III.8				41, 55	51			38

IV. Chemical Bonding

SEQUENCE	AUGUST 2002	JANUARY 2003	JUNE 2003	AUGUST 2003	JUNE 2004	AUGUST 2004	JUNE 2005	AUGUST 2005
IV.1		60		73		7, 9		†5
IV.2			12, 13	26				11
IV.3	53 a, b, c	†13	33	11, 12	9	10, †38, 70	55, 56	42
IV.4	17		24			5		13
IV.5		58			10, 39	†38		14
IV.6		14	37, 60, 61	13	11	71, 72	11	15
IV.7				34				
IV.8	12	†6		40			34	
IV.9	†60 a	59			12	41		
IV.10	23, †60 a	57			81			63
IV.11	10	10			13		12	35
IV.12	11					60		
IV.13								

V. Physical Behavior of Matter

SEQUENCE	AUGUST 2002	JANUARY 2003	JUNE 2003	AUGUST 2003	JUNE 2004	AUGUST 2004	JUNE 2005	AUGUST 2005
V.1				18				
V.2	16		16	61		11		40
V.3	51	†16, 57a, b		43		55, 56, 57	13, 39, 43	
V.4				14		12		
V.5		†16, 65, 66, †67a, b	15		14			
V.6								
V.7	7	21, 24, 52						37
V.8	48	40	14, 42	74, 76, 77, 78		13, 40, 64, 79	14, 38, 41	12, 39
V.9	36, 49	44			83	53	42	41, 43
V.10	59 e	19	23				15	
V.11								
V.12	24, 54 a, b, c			59, 60, 62	†40	42, 43	40, 68, 70	16
V.13		12		17	15, †40, 75	54	69	44, †57
V.14				16	16	15		17
V.15							16	
V.16								
V.17			64, 65		76	39	79	62
V.18				20			78	18
V.19	5	†47	17, 62				17	
V.20			18			17		45
V.21	14, 34, 50	4	22, 43	69		14		
V.22	35, 44			68	18			

SEQUENCE	AUGUST 2002	JANUARY 2003	JUNE 2003	AUGUST 2003	JUNE 2004	AUGUST 2004	JUNE 2005	AUGUST 2005
V. Physical Behavior of Matter (continued)								
V.23	28, 30	†68, †69, †70	40	15	41, 68	16		†57
V.24	33, 40	15, 30, 35, 39			17, 69		46, 80, 81	
VI. Kinetics/Equilibrium								
VI.1		5	76	35		18		
VI.2	37		75	45, 51	19, 43		45	24
VI.3	18	11			42			
VI.4	13, 60 b			22	20		18	21
VI.5	38	50		46	79	45, 63	58	
VI.6	41	38		52		19, 67	†44, 57	53, 54
VI.7		45						
VI.8			35		21	20, 64	19	55
VI.9	39	41	50	67	44		†44	46
VI.10						21	20	
VII. Organic Chemistry								
VII.1	42				22		21, 23, 82, 83	20, 22
VII.2	56 a	25, 61	21	23	23, 46	22, 75	22	19, 56
VII.3	19	49	44, 55, †56	24, 47	25, 47	44, 65	47, 60	
VII.4	55	18	25, †56	25		23	25	
VII.5								
VII.6	20	26	45		24, 63	46, 73, 74	59	47
VIII. Oxidation–Reduction								
VIII.1		46		57, 58	85	26		
VIII.2	†22	28		21		†24		
VIII.3	57 a		28	53		†24, 77	71	76
VIII.4	29	27		27				
VIII.5					48		49	
VIII.6						25		
VIII.7	†22, 57 b	4		26	26, 80	68	24	73
VIII.8				27			26	23
VIII.9		53, 54, 55	46, †63	48, 49	27	47, 76	73	
VIII.10			†63			78		74, 75
IX. Acids, Bases, and Salts								
IX.1								
IX.2	59 a		29, †48	9	28		27	25, 72
IX.3	25		31, 77	75	29	28	28	26, 49
IX.4		†34		29		27		
IX.5	59 d		30	30	30		48	82

SEQUENCE	AUGUST 2002	JANUARY 2003	JUNE 2003	AUGUST 2003	JUNE 2004	AUGUST 2004	JUNE 2005	AUGUST 2005

IX. Acids, Bases, and Salts (continued)

SEQUENCE	AUGUST 2002	JANUARY 2003	JUNE 2003	AUGUST 2003	JUNE 2004	AUGUST 2004	JUNE 2005	AUGUST 2005
IX.6	45	71, 72, †73	66		57	29, 80	76	77, 78
IX.7								27
IX.8		23, 32, †34	†48, 79		45, 56	48	50	81
IX.9	59 b, c		78	28		30	75	80

X.   Nuclear Chemistry

SEQUENCE	AUGUST 2002	JANUARY 2003	JUNE 2003	AUGUST 2003	JUNE 2004	AUGUST 2004	JUNE 2005	AUGUST 2005
X.1							62	
X.2		20	7, 39, 71, 72	50	49	31, 49, 85	29	84, 85
X.3	21	†29	34				30	29
X.4	26		32	31, 32, 70	†31, 33	83	31, 64	28, 50, 83
X.5	8	†29, 56a		33			32	
X.6	58 a, b, c							
X.7	43		49		64			
X.8		56b			32			
X.9		56c						
X.10		33	69, 70, 73	71	82			
X.11		31						

Glossary of Important Terms

absolute zero   The lowest possible temperature, written as 0 K or $-273°C$.

accuracy   The closeness of a measurement to an accepted value; see also **precision**.

acid   See **Arrhenius acid**; **Brönsted-Lowry acid**.

activated complex   The intermediate state between reactants and products in a chemical reaction; the peak of the potential energy diagram.

activation energy   The minimum energy needed to initiate a reaction.

addition polymerization   The joining of unsaturated monomers by a series of addition reactions.

addition reaction   The process in which a substance reacts across a double or triple bond in an organic compound.

alcohol   An organic compound containing a hydroxyl (—OH) group.

aldehyde   An organic compound containing a carbonyl group with at least one hydrogen atom attached to the carbonyl carbon.

alkali metal   Any Group 1 element, excluding hydrogen.

alkaline earth element   Any Group 2 element.

alkane   A hydrocarbon containing only single bonds between adjacent carbon atoms.

alkene   A hydrocarbon containing one double bond between two adjacent carbon atoms.

Reprinted with permission from *Let's Review: Chemistry* by Albert S. Tarendash, © 1998 by Barron's Educational Series, Inc.

alkyl group   An open-chained hydrocarbon less one hydrogen atom; for example, CH_3 = methyl group, C_2H_5 = ethyl group. Unspecified alkyl groups are designed by the letter R.

alkyne   A hydrocarbon containing one triple bond between two adjacent carbon atoms.

allotrope   A specific form of an element that can exist in more than one form; graphite and diamond are allotropes of the element carbon.

alloy   A solid metallic solution.

alpha decay   The radioactive process in which an alpha particle is emitted.

alpha particle (α)   A helium-4 nucleus.

amide   An organic compound containing the $CONH_2$ functional group.

amine   A hydrocarbon derivative containing an amino group.

amino acid   An organic compound containing at least one amino group and one carboxyl group.

amino group   An ammonia molecule less one hydrogen atom; $—NH_2$.

anhydrous   Pertaining to a compound from which the water of crystallization has been removed.

anode   The electrode at which oxidation occurs.

aqueous   Pertaining to a solution in which water is the solvent.

aromatic hydrocarbon   Any ring hydrocarbon whose electronic structure is related to that of benzene.

Arrhenius acid   Any substance that releases H^+ ions in water.

Arrhenius base   Any substance that releases OH^- ions in water.

atmospheric pressure   1 standard atmosphere (atm) = 1.013 kilopascals.

atom   The basic unit of an element.

atomic mass   The weighted average of the masses of the isotopes of an element.

atomic mass unit (u)   One-twelfth the mass of a carbon-12 atom.

atomic number   The number of protons in the nucleus of an atom; the atomic number defines the element.

atomic radius   A measure of the size of an atom.

Avogadro's hypothesis   Equal volumes of gases, measured at the same temperature and pressure, contain equal numbers of particles.

Avogadro's number (N_A)   The number of particles in 1 mole; 6.02×10^{23}.

battery   A commercial Voltaic cell.

benzene C_6H_6; the parent hydrocarbon of all aromatic compounds.

beta decay   The radioactive process in which a beta particle is emitted.

beta (−) particle (β^-)   An electron.

beta (+) particle (β^+)   A positron.

binary compound   A compound containing two elements.

binding energy   The energy released when a nucleus is assembled from its nucleons.

boiling   The transition of liquid to gas; boiling occurs when the vapor pressure of a liquid equals the atmospheric pressure above the liquid.

boiling point   The temperature at which boiling occurs; the temperature at which the liquid and vapor phases of a substance are in equilibrium.

boiling point elevation   The increase in the boiling point of a solvent due to the presence of solute particles.

bond energy   The energy needed to break a chemical bond.

Boyle's law   At constant temperature and mass, the pressure of an ideal gas is inversely proportional to its volume; $P_1V_1 = P_2V_2$.

breeder reactor   A fission reactor that generates its own nuclear fuel.

bright-line spectrum   The lines of visible light emitted by elements as electrons fall to lower energy levels.

Brönsted-Lowry acid   A substance that can donate H^+ ions.

Brönsted-Lowry base   A substance that can accept H^+ ions.

carbonyl group   The functional group characteristic of aldehydes and ketones; $>C{=}O$.

carboxyl group   The functional group characteristic of organic acids; —COOH.

catalyst   A substance that speeds a chemical reaction by lowering the activation energy of the reaction.

cathode   The electrode at which reduction occurs.

Celsius (C) scale   The temperature scale on which the freezing and boiling points of water (at 1 atm) are set at 0 and 100, respectively.

chain reaction   A chemical or nuclear reaction in which one step supplies energy or reactants for the next step.

Charles's law   At constant pressure and mass, the volume of an ideal gas is directly proportional to the Kelvin temperature; $\dfrac{V_1}{T_1} = \dfrac{V_2}{T_2}$

chemical bond   The stabilizing of two atoms by sharing or transferring electrons.

chemical equation   A shorthand listing of reactants, products, and molar quantities in a chemical reaction.

chemical equilibrium   The state in which the rates of the forward and reverse reactions are equal.

chemical family   See **group**.

coefficient   A number in a chemical equation that indicates how many particles of a reactant or product are required or formed in the reaction.

colligative property   A property that depends on the number of particles present rather than the type of particle; see also **boiling point elevation; freezing point depression**.

combined (ideal) gas law   At constant mass, the product of the pressure and volume divided by the Kelvin temperature is a constant;

$$\frac{P_1 V_1}{T_1} = \frac{P_2 V_2}{T_2}$$

compound   A combination of two or more elements with a fixed composition by mass.

concentrated   Pertaining to a solution that contains a relatively large quantity of solute.

concentration   The "strength" of a solution; the quantity of solute relative to the quantity of solvent.

condensation   The change from gas to liquid.

condensation polymerization   The joining of monomers by a series of dehydration reactions.

control rod   The part of a fission reactor that controls the rate of fission by absorbing neutrons.

coordinate covalent bond   A single covalent bond in which the pair of electrons is supplied by one atom.

covalent bond   A chemical bond formed by the sharing of electrons.

cracking   The process of breaking large hydrocarbon molecules into smaller ones in order to increase the yield of compounds such as gasoline.

crystal   A solid whose particles are arranged in a regularly repeating pattern.

decomposition   A reaction in which a compound forms two or more simpler substances.

density   Mass per unit volume; $d = \dfrac{m}{V}$

deposition   The direct transition from gas to solid.

deuterium   The isotope of hydrogen with a mass number of 2.

diatomic molecule   A neutral particle consisting of two atoms; Br_2 and CO are diatomic molecules.

diffusion   The movement of one substance through another.

dihydroxy alcohol   An organic compound with two hydroxyl groups.

dilute   (adjective) Pertaining to a solution that contains a relatively small quantity of solute; (verb) to reduce the concentration of a solution by adding solvent.

dipole   An unsymmetrical charge distribution in a neutral molecule.

dipole-dipole attraction   The attractive force between two oppositely charged dipoles of neighboring polar molecules.

dissociation   The separation of an ionic compound in solution into positive and negative ions.

distillation   The simultaneous boiling of a liquid and condensation of its vapor.

double bond   A covalent bond in which two pairs of electrons are shared by two adjacent atoms.

ductility   The property of a substance that allows it to be drawn into a wire; metallic substances possess ductility.

dynamic equilibrium   The state in which the rates of opposing processes are equal; see also **chemical equilibrium**; **phase equilibrium**; **solution equilibrium**.

electrochemical cell—either a *voltaic cell* **or an electrolytic cell**   A device that produces usable electrical energy from a spontaneous redox reaction; see also **battery**.

electrode   A conductor in an electrochemical or electrolytic cell that serves as the site of oxidation or reduction.

electrolysis   A nonspontaneous redox reaction driven by an external source of electricity.

electrolyte   A substance whose aqueous solution conducts electricity.

electrolytic cell   A device for carrying out electrolysis.

electron   The elementary unit of negative charge.

electron-dot diagram   See **Lewis structure**.

electronegativity   The measure of an atom's attraction for a bonded pair of electrons.

electroplating   The use of an electric current to deposit a layer of metal on a negatively charged object.

element   A substance all of whose atoms have the same atomic number.

empirical formula   A formula in which the elements are present in the smallest whole-number ratio; NO_2 is an empirical formula, but C_2H_4 is not.

endothermic reaction   A reaction that absorbs energy; ΔH is positive for an endothermic reaction.

end point   The point in a titration that signals that equivalent quantities of reactants have been added.

energy   A quantity related to an object's capacity to do work.

enthalpy change (ΔH)   The heat energy absorbed or released by a system at constant pressure.

entropy (S)   The measure of the randomness or disorder of a system.

entropy change (ΔS)   An increase or decrease in the randomness of a system.

equilibrium   See **dynamic equilibrium**.

ester   The organic product of esterification.

esterification   The reaction of an acid with an alcohol to produce an ester and water.

ethanoic acid CH_3COOH; acetic acid.

ethanol CH_3CH_2OH; ethyl (grain) alcohol.

ethene C_2H_4; ethylene; the parent of the alkene family of hydrocarbons.

ether   An organic compound containing the arrangement R—O—R.

ethyne C_2H_2; acetylene; the parent of the alkyne family of hydrocarbons.

evaporation   The surface transition of liquid to gas.

excited state   A condition in which one or more electrons in an atom are no longer in the lowest possible energy state.

exothermic reaction   A reaction that releases energy; ΔH is negative for an exothermic reaction.

fermentation   The (anaerobic) oxidation of a sugar such as glucose to produce ethanol and carbon dioxide; the reaction is catalyzed by enzymes.

filtration   A method of separating a liquid from the particles suspended in it.

first ionization energy   The quantity of energy needed to remove the most loosely held electron from an isolated neutral atom.

fission   A nuclear reaction in which a heavy nuclide splits to form lighter nuclides and energy.

fission reactor   A device for producing electrical energy by means of a controlled fission reaction.

formula mass   The sum of the masses of the atoms in a formula; units are atomic mass units (u) or grams per mole (g/mol).

fractional distillation   The separation of organic substances based on differences in their boiling points.

freezing   The transition from liquid to solid.

freezing point   The temperature at which freezing occurs.

freezing point depression (lowering)   The decrease in the freezing point of a solvent due to the presence of solute particles.

fuel rod   The part of a nuclear reactor that contains the fissionable material.

functional group   An atom or group of atoms that confers specific properties on an organic molecule.

fusion   A synonym for *melting*; also, a nuclear process in which light nuclides join to form heavier nuclides and produce radiant energy.

fusion reactor   An experimental device for producing a controlled fusion reaction and generating electrical energy from it.

gas   The phase in which matter has neither definite shape nor definite volume.

gram-atomic mass   The molar mass of an element expressed in grams per mole (g/mol).

gram-formula mass   See **molar mass**.

gram-molecular mass   The molar mass of a molecule.

ground state   The electron configuration of an atom in the lowest energy state.

group   The elements within a single vertical column of the Periodic Table.

half-cell   The part of an electrochemical cell in which oxidation or reduction occurs.

half-life   The time needed for a substance to decay to one-half its initial mass.

half-reaction   The oxidation or reduction portion of a redox reaction.

halogen   An element in Group 17 of the Periodic Table; F, Cl, Br, I, At.

heat energy   The energy released or absorbed by a system undergoing a change in temperature, in phase, or in composition.

heat of fusion (H_f)   The heat energy absorbed when a unit mass of solid changes to liquid at its melting point; $H_{f(ice)} = 80$ calories per gram.

heat of reaction (ΔH)   The heat energy absorbed or released as a result of a chemical reaction.

heat of vaporization (H_v)   The heat energy absorbed when a unit mass of liquid changes to gas at its boiling point; $H_{v(water)} = 540$ calories per gram.

heavy water   A molecule of water in which the hydrogen atoms have a mass number of 2; deuterium oxide.

heterogeneous mixture   A nonuniform mixture.

homogeneous mixture   A mixture with a uniform distribution of particles; a solution is one example of a homogeneous mixture.

homologous series   A group of organic compounds with related structures and properties; each successive member of the series differs from the one before it by a specific number of carbon and hydrogen atoms (usually CH_2).

hydrate   A crystalline compound that has water molecules incorporated into its crystal structure; common examples include $CuSO_4 \cdot 5H_2O$ and $Na_2SO_4 \cdot 10H_2O$ [also written as $CuSO_4(H_2O)_5$ and $Na_2SO_4(H_2O)_{10}$].

hydration   The association of water molecules with an ion or another molecule.

hydride   A binary compound of an active metal and hydrogen; the oxidation state of hydrogen is -1.

hydrocarbon   An organic compound composed of carbon and hydrogen.

hydrogen bond   An unusually strong intermolecular attraction that results when hydrogen is bonded to a small, highly electronegative atom such as F, O, or N.

hydrolysis A reaction in which a water molecule breaks a chemical bond; the reaction between certain salts and water to produce an excess of hydronium or hydroxide ions.

hydronium ion H_3O^+; the conjugate acid of H_2O; responsible for acidic properties in water solutions.

hydroxide ion OH^-; the conjugate base of H_2O; responsible for basic properties in water solutions.

ideal gas A model of a gas in which the particles have no volume, do not attract or repel each other, and collide without loss of energy; real gases approximate ideal gas behavior under conditions of low pressure and high temperature.

ideal gas law The relationship obeyed by an ideal gas; see **combined (ideal) gas law**.

indicator A substance that undergoes a color change to signal a change in chemical conditions; acid-base indicators change color over specified pH ranges.

inert (noble) gas An element in Group 18 of the Periodic Table; Ne, Ar, Kr, Xe, Rn. (He is also associated with Group 18.)

inorganic compound A compound that is not a hydrocarbon derivative.

ion A particle in which the numbers of protons and electrons are not equal.

ion-dipole attraction The attractive force between an ion and the oppositely charged dipole of a neighboring polar molecule.

ionic bond The electrostatic attraction of positive and negative ions in an ionic compound; an electronegativity difference of 1.7 or greater indicates the presence of an ionic bond.

ionic compound A substance whose particles consist of positive and negative ions.

ionization energy The quantity of energy needed to remove an electron from an atom or ion; see also **first ionization energy**.

isomers Different compounds that have the same molecular formula.

isotopes Atoms having the same atomic number but different mass numbers; atoms of the same element with differing numbers of neutrons.

IUPAC International Union of Pure and Applied Chemistry; the scientific group responsible for all major policies in chemistry, including the naming of elements and compounds.

joule (J)   The unit of work and energy in the SI (metric) system.

Kelvin (K)   A measure of absolute temperature; the Kelvin scale begins at 0 and is related to the Celsius scale by the equation $K = C + 273$; a temperature *difference* of 1 K is equal to a temperature *difference* of 1 C°.

ketone   An organic compound containing a carbonyl group with no hydrogen atoms directly attached to the carbonyl carbon.

kilo-   The metric prefix signifying 1,000.

kilojoule (kJ)   1,000 joules.

kinetic energy   The energy associated with the motion of an object.

kinetic-molecular theory (KMT)   The theory that explains the structure and behavior of idealized models of gases, liquids, and solids.

Le Châtelier's principle   When a system at equilibrium is subjected to a stress, the system will shift in order to lessen the effects of the stress. Eventually, a new equilibrium point is established.

Lewis structure   A shorthand notation for illustrating the ground-state valence electron configuration of an atom.

liquid   The phase in which matter has a definite volume but an indefinite shape; a liquid takes the shape of its container.

liter (L)   A unit of volume in the metric system; 1 liter = 1,000 cubic centimeters; 1 liter = 1 cubic decimeter; 1 liter is approximately equal to 1 quart.

litmus   An acid-base indicator that is red in acidic solutions and blue in basic solutions.

macromolecule   A giant molecule formed by network bonding or by polymerization.

malleability   The property by which a substance is able to be formed into various shapes; metallic substances possess malleability.

mass number   The number of nucleons in a nuclide.

melting   The transition from solid to liquid.

melting point   The temperature at which the vapor pressure of a solid equals the vapor pressure of the liquid; the temperature at which the solid and liquid phases of a substance are in equilibrium; see also **freezing point**.

meniscus   The curved surface of a liquid, caused by the attraction of the particles of the liquid and the container holding the liquid (e.g., water in a graduated cylinder), or by the mutual attraction of the particles of the liquid (e.g., mercury).

metal   A substance composed of atoms with low ionization energies and relatively vacant valence levels; Na, Fe, Ag, and Ba are metallic substances.

metallic bond   The delocalization of the valence electrons among the kernels of the metal atoms; "mobile valence electrons immersed in a sea of positive ions."

metalloid   An element that has both metallic and nonmetallic properties; examples of metalloids include B, Ge, Si, and Te.

methanal   HCHO; formaldehyde; the simplest aldehyde.

methane CH_4; the parent of the alkane family of hydrocarbons.

methanoic acid   HCOOH; formic acid; the simplest organic acid.

methanol CH_3OH; methyl (wood) alcohol; the simplest alcohol.

milli-   The metric prefix signifying 1/1,000.

milliliter (mL)   1/1,000 liter; 1 liter = 1,000 milliliters.

miscible   A solution of liquids that is soluble in all proportions; ethanol and water are a miscible pair of liquids.

mixture   A material consisting of two or more components and having a variable composition.

moderator   A substance used to produce slow neutrons and promote nuclear fission; graphite and heavy water are used as moderators in fission reactors.

molar mass   The mass of any atom, element, ion, or compound expressed in grams per mole (g/mol).

molarity   The concentration of a solution, measured as the number of moles of solute per liter of solution.

molar volume   The volume occupied by 1 mole of an ideal gas; 22.4 liters at STP.

mole   The number of atoms contained in 12 grams of carbon-12; see also **Avogadro's number**.

molecular formula   A chemical formula that lists the number of atoms present but does not show the arrangement of the atoms in space.

molecular mass   The sum of the masses of the atoms in a molecule; units are atomic mass units (u) or grams per mole (g/mol).

molecule   The smallest unit of a nonionic substance; Ar, Cl_2, and NH_3 are molecules.

monatomic molecule   A molecule consisting of one atom; Xe and He are monatomic molecules.

monomer   The basic unit of a polymer; the monomer of a protein is an amino acid; the monomer of starch is glucose.

network solid   A substance formed by a two- or three-dimensional web of covalent bonds to produce a macromolecule; diamond and SiO_2 are network solids.

neutralization   The reaction of equivalent amounts of hydronium and hydroxide ions in aqueous solution; the principal product is water. When the water is evaporated, the spectator ions form a salt; see also **spectator ion**.

neutron   A neutral nuclear particle with a mass comparable to that of a proton.

noble gas   See **inert (noble) gas**.

nonelectrolyte   A substance whose aqueous solution does not conduct electricity; glucose is a nonelectrolyte.

nonmetal   A substance that does not have characteristic metallic properties; C and S are nonmetallic elements.

nonpolar bond   A covalent bond in which the electron pair or pairs are shared equally by both atoms.

nonpolar molecule   A molecule containing only nonpolar bonds, such as N_2, or a molecule with a symmetrical charge distribution, such as CCl_4 and CO_2.

normal boiling point   The boiling temperature of a substance at a pressure of 1 atmosphere.

nuclear equation   A shorthand listing of reactant and product nuclides in a nuclear reaction.

nucleon   A constituent of an atomic nucleus; a proton or a neutron.

nucleus   The portion of the atom that contains more than 99.9 percent of the atom's mass; the nucleus is small, dense, and positively charged.

ore   A native mineral from which a metal or metals can be extracted.

organic chemistry   The study of the hydrocarbons and their derivatives; the chemistry of carbon.

organic compound A compound that is a hydrocarbon or a hydrocarbon derivative.

oxidation The loss or apparent loss of electrons in a chemical reaction.

oxidation number (state) The charge that an atom has or appears to have when certain arbitrary rules are applied; oxidation numbers are useful for identifying the atoms that are oxidized and reduced in a redox reaction.

oxidizing agent The particle in a redox reaction that causes another particle to be oxidized; as a result, an oxidizing agent is reduced.

paraffin A common name for a mixture of solid alkanes; another name for paraffin is wax.

percent composition by mass The number of grams of an element (or group of elements) present in 100 grams of an ion or compound.

period One of the horizontal rows of the Periodic Table; the period number indicates the valence level of an element.

periodic law The properties of elements recur at regular intervals and depend on their nuclear charges; "The properties of elements are a periodic function of their atomic numbers."

peroxide A compound in which oxygen has an oxidation state of -1; H_2O_2 and BaO_2 are peroxides.

petroleum Crude oil containing a mixture of hydrocarbons.

pH A scale of acidity and basicity based on the hydronium ion concentration in aqueous solution; $pH = -\log[H_3O^+]$. At 298 K, a pH less than 7 indicates an acidic solution; a pH greater than 7, a basic solution; a pH of 7, a neutral solution.

phase equilibrium The state in which the rates of opposing phase changes (freezing–melting, boiling–condensation, sublimation–deposition) are equal.

phenolphthalein An acid–base indicator that is colorless in acidic solutions and pink in basic solutions.

photon A fundamental unit of radiant energy; a quantum of radiation.

polar bond A covalent bond in which the electron pair or pairs are shared unequally by both atoms; the atom with the larger electronegativity has more of the electron density surrounding it.

polar molecule A molecule with an unsymmetrical charge distribution, such as H_2O; a dipole.

polyatomic ion   An ion composed of more than one atom; SO_4^{2-} is a polyatomic ion.

polymer   A macromolecule consisting of a chain of simpler units; polyethylene is a polymer of ethene.

polymerization   See **polymer**; **addition polymerization**; **condensation polymerization**.

positron   A positively charged electron; a particle of antimatter.

potential energy   The energy associated with the position of an object; a "stored" form of energy.

precipitate   A deposit formed by the appearance of an excess of solid solute in a saturated solution.

precision   The closeness of a series of measurements to one another; see also **accuracy**.

pressure   The force exerted on an object divided by the surface area of the object; $P = \dfrac{F}{A}$.

principal energy level   An integer beginning with 1 that describes the approximate distance of an electron from the nucleus of an atom.

product(s)   The substance or substances that are formed in a chemical process; products are on the right side of a chemical equation.

propanone $(CH_3)_2CO$; acetone; the simplest ketone.

proton   A nuclear particle with a positive charge equal to the negative charge on the electron; a nucleon.

radiant energy   Electromagnetic energy; visible light and X rays are examples of radiant energy.

radioactive dating   The use of radioactive isotopes to measure the age of an object.

radioactive tracer   A radioisotope used to indicate the path of an atom in a chemical reaction.

radioactivity   The spontaneous breakdown of a radioactive nuclide.

radioisotope   A radioactive isotope.

reactant(s)   The substance or substances that react in a chemical process; reactants are on the left side of a chemical equation.

redox reaction   A chemical reaction in which oxidation-reduction takes place.

reducing agent   The particle in a redox reaction that causes another particle to be reduced; as a result, a reducing agent is oxidized.

reduction   The gain or apparent gain of electrons in a chemical reaction.

salt   The spectator-ion product of a neutralization reaction; see also **spectator ion**.

salt bridge   A device for allowing the flow of ions in an electrochemical cell.

saponification   The reaction of an ester with a base to produce an alcohol and the sodium salt of an organic acid; soap is produced by saponifying fats with NaOH.

saturated hydrocarbon   A hydrocarbon containing only single carbon-carbon bonds.

saturated solution   A solution in which the pure solute is in equilibrium with the dissolved solute; a solution that contains the maximum amount of dissolved solute.

significant digit(s) [figure(s)]   The number or numbers that are part of a measurement. If there are two or more, all but the last figure is known; the last figure is the experimenter's best estimate.

single bond   A covalent bond in which one pair of electrons is shared by two adjacent atoms.

solid   The phase in which matter has both definite shape and definite volume.

solubility   The amount of solute needed to produce a saturated solution with a given amount of solvent.

solute(s)   The substance or substances dissolved in a solution.

solution   A homogeneous mixture whose particles are extremely small.

solution equilibrium   The state in which the undissolved and dissolved solutes are in dynamic equilibrium; in a solid-liquid solution, the rate of dissolving equals the rate of crystallization.

solvent   The substance in which the solute is dissolved.

spectator ion   An ion that does not take part in a chemical reaction; in the (acid-base) reaction between NaOH(aq) and HCl(aq), Na^+(aq) and Cl^-(aq) are spectator ions; in the (redox) reaction between Zn(s) and $Cu(NO_3)_2$(aq), NO_3^-(aq) is the spectator ion.

standard solution   A solution whose concentration is accurately known; a standard solution is used for analyzing other substances.

standard temperature and pressure (STP)   273 K and 1 atmosphere.

Stock system   A systematic method of naming chemical compounds in which the *positive* oxidation number is written as a Roman numeral in parentheses after the element. For example, the Stock name of the compound Fe_2O_3 is iron(III) oxide.

stoichiometry   The study of quantitative relationships in substances and reactions; chemical mathematics.

strong acid   In aqueous solution, a substance that ionizes almost completely to hydronium ion.

strong base   In aqueous solution, a substance that ionizes or dissociates almost completely to hydroxide ion.

structural formula   A chemical formula that illustrates the spatial arrangement of each atom.

sublimation   The direct transition from solid to gas; $CO_2(s)$ and $I_2(s)$ sublime at atmospheric pressure.

substance   An element or a compound.

supersaturated solution   A solution that contains more dissolved solute than a saturated solution at the same temperature; supersaturation is an unstable condition.

temperature   A measure of the average kinetic energy of the particles of a substance.

titration   The addition of a known volume of a standard solution in order to determine the concentration of an unknown solution.

toluene C_7H_8; methylbenzene.

transition element   An element whose atoms contain unfilled d sublevels; an element in Groups 3–11 of the Periodic Table.

transmutation   The conversion of one element to another by a nuclear process.

triple bond   A covalent bond in which three pairs of electrons are shared by two adjacent atoms.

tritium   The radioactive isotope of hydrogen with a mass number of 3.

unsaturated hydrocarbon   A hydrocarbon containing double and/or triple carbon-carbon bonds.

unsaturated solution   A solution that contains less dissolved solute than a saturated solution at the same temperature.

valence electron   An electron in the outermost principle energy level.

Voltaic cell   A device that produces usable electrical energy from a spontaneous redox reaction; see also **battery**.

water of hydration   The water molecules that are part of the crystalline structure of certain compounds.

weak acid   In aqueous solution, a substance that is poorly ionized and produces only a small concentration of hydronium ion.

weak base   In aqueous solution, a substance that is poorly ionized or dissociated and produces only a small concentration of hydroxide ion.

Using the Chemistry Reference Tables

Table A—Standard Temperature and Pressure

Table A provides the values for standard pressure (in kPa and atm) and standard temperature (in K and °C). These values are useful in solving gas-law problems.

STANDARD TEMPERATURE AND PRESSURE

Name	Value	Unit
Standard Pressure	101.3 kPa	kilopascal
	1 atm	atmosphere
Standard Temperature	273 K	kelvin
	0°C	degree Celsius

Table B—Physical Constants for Water

Table B provides the values for the *heat of fusion of ice*, the *heat of vaporization of water*, and the *specific heat capacity for liquid water*. These constants are used frequently in the solution of problems involving heat transfer and phase changes.

PHYSICAL CONSTANTS FOR WATER

Heat of Fusion	333.6 J/g
Heat of Vaporization	2259 J/g
Specific Heat Capacity of H_2O (ℓ)	4.2 J/g•K

Table C—Selected Prefixes

Table C provides a number of the more important selected metric prefixes, their factors and symbols. A prefix placed in front of a base unit creates a multiple or submultiple of that unit: For example, *deci*liter (dL) is a unit of volume whose value is 10^{-1} liter.

SELECTED PREFIXES

Factor	Prefix	Symbol
10^3	kilo-	k
10^{-1}	deci-	d
10^{-2}	centi-	c
10^{-3}	milli-	m
10^{-6}	micro-	μ
10^{-9}	nano-	n
10^{-12}	pico-	p

Table D—Selected Units

Table D provides the names and symbols for a number of quantities that are used in chemistry. For example, heat, work, and energy are all measured in *joules*, a unit that has the symbol *J*.

SELECTED UNITS

Symbol	Name	Quantity
m	meter	length
kg	kilogram	mass
Pa	pascal	pressure
K	kelvin	temperature
mol	mole	amount of substance
J	joule	energy, work, quantity of heat
s	second	time
L	liter	volume
ppm	part per million	concentration
M	molarity	solution concentration

Table E—Selected Polyatomic Ions

Table E provides a listing of the names, formulas, and charges of various polyatomic ions. The Table is very useful for writing the correct formulas of compounds that have polyatomic ions as part of their structure.

SELECTED POLYATOMIC IONS

H_3O^+	hydronium	CrO_4^{2-}	chromate
Hg_2^{2+}	dimercury (I)	$Cr_2O_7^{2-}$	dichromate
NH_4^+	ammonium	MnO_4^-	permanganate
$C_2H_3O_2^-$ CH_3COO^- } acetate		NO_2^-	nitrite
		NO_3^-	nitrate
CN^-	cyanide	O_2^{2-}	peroxide
CO_3^{2-}	carbonate	OH^-	hydroxide
HCO_3^-	hydrogen carbonate	PO_4^{3-}	phosphate
$C_2O_4^{2-}$	oxalate	SCN^-	thiocyanate
ClO^-	hypochlorite	SO_3^{2-}	sulfite
ClO_2^-	chlorite	SO_4^{2-}	sulfate
ClO_3^-	chlorate	HSO_4^-	hydrogen sulfate
ClO_4^-	perchlorate	$S_2O_3^{2-}$	thiosulfate

Table F—Solubility Guidelines (for Aqueous Solutions at 25°C)

Table F provides a number of general rules that enable one to determine the solubilities of various *ionic* compounds in water at 25°C. The table is split between those ions that form *soluble* compounds and those ions that form *insoluble* compounds. The table also lists a number of notable exceptions to these rules. For example, ammonium carbonate is soluble in water, while calcium phosphate is not.

SOLUBILITY GUIDELINES

Ions That Form Soluble Compounds	Exceptions	Ions That Form Insoluble Compounds	Exceptions
Group 1 ions (Li$^+$, Na$^+$, etc.)		carbonate (CO$_3$$^{2-}$)	when combined with Group 1 ions or ammonium (NH$_4$$^+$)
ammonium (NH$_4$$^+$)		chromate (CrO$_4$$^{2-}$)	when combined with Group 1 ions or ammonium (NH$_4$$^+$)
nitrate (NO$_3$$^-$)		phosphate (PO$_4$$^{3-}$)	when combined with Group 1 ions or ammonium (NH$_4$$^+$)
acetate (C$_2$H$_3$O$_2$$^-$ or CH$_3$COO$^-$)		sulfide (S^{2-})	when combined with Group 1 ions or ammonium (NH$_4$$^+$)
hydrogen carbonate (HCO$_3$$^-$)		hydroxide (OH$^-$)	when combined with Group 1 ions, Ca^{2+}, Ba^{2+}, or Sr^{2+}
chlorate (ClO$_3$$^-$)			
perchlorate (ClO$_4$$^-$)			
halides (Cl$^-$, Br$^-$, I$^-$)	when combined with Ag$^+$, Pb^{2+}, and Hg$_2$$^{2+}$		
sulfates (SO$_4$$^{2-}$)	when combined with Ag$^+$, Ca^{2+}, Sr^{2+}, Ba^{2+}, and Pb^{2+}		

Table G—Solubility Curves

Table G presents solubility curves for a number of solids and gases dissolved in water over the temperature range 0°C–100°C. The curves can be used to calculate the solubility of a substance in water at a given temperature, calculate the effect of raising or lowering the temperature, describe the general trend in the solubility of a substance with temperature changes, or compare the solubilities of various substances in water.

SOLUBILITY CURVES

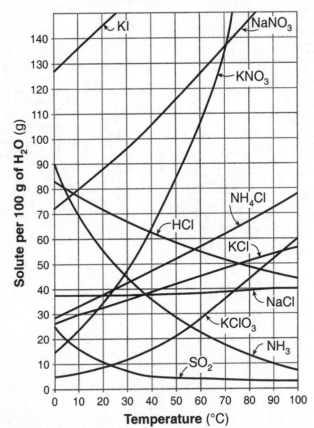

Table H—Vapor Pressure of Four Liquids

Table H presents four curves, which trace the vapor pressures of propanone (acetone), ethanol, water, and ethanoic (acetic) acid between 0°C and 100°C. In each case, the vapor pressure rises slowly at low temperatures and climbs more rapidly at higher temperatures. The curves allow one to calculate the boiling point of one of the liquids at a specific (external) pressure. For example, the boiling point of propanone at an external pressure of 70 kPa is 45°C. Standard atmospheric pressure (101.3 kPa) is indicated by a dashed line. The temperature at which a curve intersects this line in known as the *normal boiling point* of the liquid. For example, the normal boiling point of ethanoic acid is 118°C.

VAPOR PRESSURE OF FOUR LIQUIDS

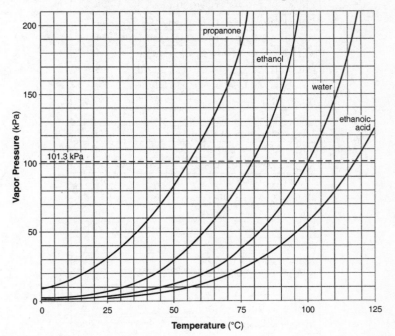

Table I—Heats of Reaction at 101.3 kPa and 298K

Table I provides a listing of various reactions and the heat energy absorbed or released (in kilojoules) when the reaction occurs. The *coefficients* of the equations indicate the numbers of *moles* of reactants and products taking part in the reactions. This table might be used to select an exothermic or endothermic reaction or calculate the heat of reaction when different molar quantities of reactants or products are given. For example, when 2 moles of $N_2(g)$ combine with 4 moles of $O_2(g)$ to produce 4 moles of $NO_2(g)$ at 101.3 kPa and 298K, 132.8 kJ $(2 \times 66.4 \text{ kJ})$ of heat will be absorbed.

HEATS OF REACTION AT 101.3 kPa AND 298K

Reaction	ΔH (kJ)*
$CH_4(g) + 2O_2(g) \longrightarrow CO_2(g) + 2H_2O(\ell)$	−890.4
$C_3H_8(g) + 5O_2(g) \longrightarrow 3CO_2(g) + 4H_2O(\ell)$	−2219.2
$2C_8H_{18}(\ell) + 25O_2(g) \longrightarrow 16CO_2(g) + 18H_2O(\ell)$	−10943
$2CH_3OH(\ell) + 3O_2(g) \longrightarrow 2CO_2(g) + 4H_2O(\ell)$	−1452
$C_2H_5OH(\ell) + 3O_2(g) \longrightarrow 2CO_2(g) + 3H_2O(\ell)$	−1367
$C_6H_{12}O_6(s) + 6O_2(g) \longrightarrow 6CO_2(g) + 6H_2O(\ell)$	−2804
$2CO(g) + O_2(g) \longrightarrow 2CO_2(g)$	−566.0
$C(s) + O_2(g) \longrightarrow CO_2(g)$	−393.5
$4Al(s) + 3O_2(g) \longrightarrow 2Al_2O_3(s)$	−3351
$N_2(g) + O_2(g) \longrightarrow 2NO(g)$	+182.6
$N_2(g) + 2O_2(g) \longrightarrow 2NO_2(g)$	+66.4
$2H_2(g) + O_2(g) \longrightarrow 2H_2O(g)$	−483.6
$2H_2(g) + O_2(g) \longrightarrow 2H_2O(\ell)$	−571.6
$N_2(g) + 3H_2(g) \longrightarrow 2NH_3(g)$	−91.8
$2C(s) + 3H_2(g) \longrightarrow C_2H_6(g)$	−84.0
$2C(s) + 2H_2(g) \longrightarrow C_2H_4(g)$	+52.4
$2C(s) + H_2(g) \longrightarrow C_2H_2(g)$	+227.4
$H_2(g) + I_2(g) \longrightarrow 2HI(g)$	+53.0
$KNO_3(s) \xrightarrow{H_2O} K^+(aq) + NO_3^-(aq)$	+34.89
$NaOH(s) \xrightarrow{H_2O} Na^+(aq) + OH^-(aq)$	−44.51
$NH_4Cl(s) \xrightarrow{H_2O} NH_4^+(aq) + Cl^-(aq)$	+14.78
$NH_4NO_3(s) \xrightarrow{H_2O} NH_4^+(aq) + NO_3^-(aq)$	+25.69
$NaCl(s) \xrightarrow{H_2O} Na^+(aq) + Cl^-(aq)$	+3.88
$LiBr(s) \xrightarrow{H_2O} Li^+(aq) + Br^-(aq)$	−48.83
$H^+(aq) + OH^-(aq) \longrightarrow H_2O(\ell)$	−55.8

*Minus sign indicates an exothermic reaction.

Table J—Activity Series

Table J lists selected metals in order of their ability to be *oxidized* in aqueous solution. For example, Zn is more easily oxidized than Cu and, as a result, Zn(s) will *replace* Cu^{2+} ions from an aqueous solution $[Zn(s) + Cu^{2+}(aq) \rightarrow Zn^{2=}(aq) + Cu(s)]$.

Table J also lists the halogens in order of their ability to be *reduced* in aqueous solution. For example, Cl_2 is more easily reduced than Br_2 and, as a result, $Cl_2(g)$ will *replace* Br^- ions from an aqueous solution $[Cl_2(g) + Br^-(aq) \rightarrow Cl^-(aq) + Br_2(\ell)]$.

ACTIVITY SERIES**

Most	Metals	Nonmetals	Most
	Li	F_2	
	Rb	Cl_2	
	K	Br_2	
	Cs	I_2	
	Ba		
	Sr		
	Ca		
	Na		
	Mg		
	Al		
	Ti		
	Mn		
	Zn		
	Cr		
	Fe		
	Co		
	Ni		
	Sn		
	Pb		
	**H_2		
	Cu		
	Ag		
	Au		
Least			Least

**Activity Series based on hydrogen standard

Table K—Common Acids

Table K provides a listing of common acids, along with their formulas.

COMMON ACIDS

Formula	Name
$HCl(aq)$	hydrochloric acid
$HNO_3(aq)$	nitric acid
$H_2SO_4(aq)$	sulfuric acid
$H_3PO_4(aq)$	phosphoric acid
$H_2CO_3(aq)$ or $CO_2(aq)$	carbonic acid
$CH_3COOH(aq)$ or $HC_2H_3O_2(aq)$	ethanoic acid (acetic acid)

Table L—Common Bases

Table L provides a listing of common hydroxide bases and aqueous ammonia, along with their formulas.

COMMON BASES

Formula	Name
$NaOH(aq)$	sodium hydroxide
$KOH(aq)$	potassium hydroxide
$Ca(OH)_2(aq)$	calcium hydroxide
$NH_3(aq)$	aqueous ammonia

Table M—Common Acid–Base Indicators

Table M provides a listing of common acid–base indicators, along with their pH ranges and their color changes. For example, the indicator thymol blue will be *yellow* at a pH value of 8.0 or less. Between the pH values 8.0 and 9.6, the color of the indicator will change gradually and will be *blue* at a pH value of 9.6 and above.

COMMON ACID–BASE INDICATORS

Indicator	Approximate pH Range for Color Change	Color Change
methyl orange	3.2–4.4	red to yellow
bromthymol blue	6.0–7.6	yellow to blue
phenolphthalein	8.2–10	colorless to pink
litmus	5.5–8.2	red to blue
bromcresol green	3.8–5.4	yellow to blue
thymol blue	8.0–9.6	yellow to blue

Table N—Selected Radioisotopes

Table N provides a listing of various radioactive nuclides, their half-lives, and their modes of decay. The Table can be used to work out half-life problems, indicate the particle formed during decay, or predict the identity of the changed nucleus after the decay has occurred.

SELECTED RADIOISOTOPES

Nuclide	Half-Life	Decay Mode	Nuclide Name
^{198}Au	2.69 d	β^-	gold-198
^{14}C	5730 y	β^-	carbon-14
^{37}Ca	175 ms	β^+	calcium-37
^{60}Co	5.26 y	β^-	cobalt-60
^{137}Cs	30.23 y	β^-	cesium-137
^{53}Fe	8.51 min	β^+	iron-53
^{220}Fr	27.5 s	α	francium-220
^{3}H	12.26 y	β^-	hydrogen-3
^{131}I	8.07 d	β^-	iodine-131
^{37}K	1.23 s	β^+	potassium-37
^{42}K	12.4 h	β^-	potassium-42
^{85}Kr	10.76 y	β^-	krypton-85
^{16}N	7.2 s	β^-	nitrogen-16
^{19}Ne	17.2 s	β^+	neon-19
^{32}P	14.3 d	β^-	phosphorus-32
^{239}Pu	2.44×10^4 y	α	plutonium-239
^{226}Ra	1600 y	α	radium-226
^{222}Rn	3.82 d	α	radon-222
^{90}Sr	28.1 y	β^-	strontium-90
^{99}Tc	2.13×10^5 y	β^-	technetium-99
^{232}Th	1.4×10^{10} y	α	thorium-232
^{233}U	1.62×10^5 y	α	uranium-233
^{235}U	7.1×10^8 y	α	uranium-235
^{238}U	4.51×10^9 y	α	uranium-238

ms = milliseconds; s = seconds; min = minutes;
h = hours; d = days; y = years

Table O—Symbols Used in Nuclear Chemistry

Table O provides a listing of the common particles appearing in nuclear equations, together with their symbols. The Table may be used in conjunction with *Table N*. It might be used to identify the nuclear charge or mass number of a particular particle.

SYMBOLS USED IN NUCLEAR CHEMISTRY

Name	Notation	Symbol
alpha particle	^4_2He or $^4_2\alpha$	α
beta particle (electron)	$^0_{-1}\text{e}$ or $^0_{-1}\beta$	β^-
gamma radiation	$^0_0\gamma$	γ
neutron	^1_0n	n
proton	^1_1H or ^1_1p	p
positron	$^0_{+1}\text{e}$ or $^0_{+1}\beta$	β^+

Table P—Organic Prefixes

Table P provides a listing of the prefixes used with organic compounds having 1 to 10 carbon atoms. For example, the compound whose name is 2-*pent*anone has 5 carbon atoms in its structure.

ORGANIC PREFIXES

Prefix	Number of Carbon Atoms
meth-	1
eth-	2
prop-	3
but-	4
pent-	5
hex-	6
hept-	7
oct-	8
non-	9
dec-	10

Table Q—Homologous Series of Hydrocarbons

Table Q provides a listing of the names and general formulas for hydrocarbons. An alkane contains *only* single bonds between its carbon atoms; an alkene contains *one* double bond between two of its carbon atoms, and an alkyne contains *one* triple bond between two of its carbon atoms. An example (name and structural formula) is provided for each homologous series.

HOMOLOGOUS SERIES OF HYDROCARBONS

Name	General Formula	Examples	
		Name	Structural Formula
alkanes	C_nH_{2n+2}	ethane	H H | | H—C—C—H | | H H
alkenes	C_nH_{2n}	ethene	H H \ / C=C / \ H H
alkynes	C_nH_{2n-2}	ethyne	H—C≡C—H

n = number of carbon atoms

Table R—Organic Functional Groups

Table R provides a listing of the common organic functional groups that enable one to classify an organic compound containing elements other than carbon and hydrogen. Along with the name of the class of compound, the formula of the functional group is given, as well as a general formula for that class. In addition, one example (name and condensed structural formula) is given for each class of organic compound.

ORGANIC FUNCTIONAL GROUPS

Class of Compound	Functional Group	General Formula	Example
halide (halocarbon)	—F (fluoro-) —Cl (chloro-) —Br (bromo-) —I (iodo-)	$R-X$ (X represents any halogen)	$CH_3CHClCH_3$ 2-chloropropane
alcohol	—OH	$R-OH$	$CH_3CH_2CH_2OH$ 1-propanol
ether	—O—	$R-O-R'$	$CH_3OCH_2CH_3$ methyl ethyl ether
aldehyde	$\overset{\displaystyle O}{\overset{\|}{-C-H}}$	$R-\overset{\displaystyle O}{\overset{\|}{C}}-H$	$CH_3CH_2\overset{\displaystyle O}{\overset{\|}{C}}-H$ propanal
ketone	$\overset{\displaystyle O}{\overset{\|}{-C-}}$	$R-\overset{\displaystyle O}{\overset{\|}{C}}-R'$	$CH_3\overset{\displaystyle O}{\overset{\|}{C}}CH_2CH_2CH_3$ 2-pentanone
organic acid	$\overset{\displaystyle O}{\overset{\|}{-C-OH}}$	$R-\overset{\displaystyle O}{\overset{\|}{C}}-OH$	$CH_3CH_2\overset{\displaystyle O}{\overset{\|}{C}}-OH$ propanoic acid
ester	$\overset{\displaystyle O}{\overset{\|}{-C-O-}}$	$R-\overset{\displaystyle O}{\overset{\|}{C}}-O-R'$	$CH_3CH_2\overset{\displaystyle O}{\overset{\|}{C}}OCH_3$ methyl propanoate
amine	$-\overset{\|}{N}-$	$R-\overset{R'}{\overset{\|}{N}}-R''$	$CH_3CH_2CH_2NH_2$ 1-propanamine
amide	$-\overset{\displaystyle O}{\overset{\|}{C}}-\overset{\|}{N}H$	$R-\overset{\displaystyle O}{\overset{\|}{C}}-\overset{R'}{\overset{\|}{N}}H$	$CH_3CH_2\overset{\displaystyle O}{\overset{\|}{C}}-NH_2$ propanamide

R represents a bonded atom or group of atoms.

Periodic Table of the Elements

The *Periodic Table of the Elements* lists the known elements along with information about atomic number, atomic mass, symbol, ground state electron configuration, and selected oxidation states. In addition, the Group and Period numbers for each element are listed at the top and left side of the Periodic Table, respectively. At the bottom of the

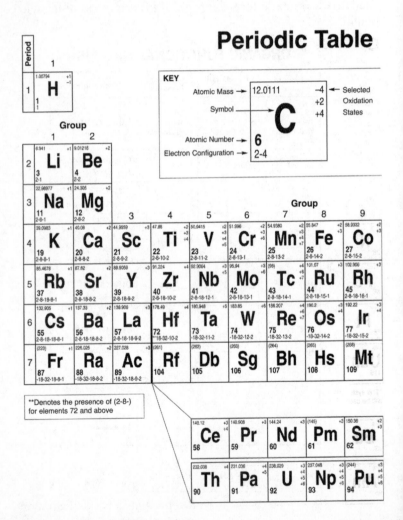

Periodic Table, the lanthanide (elements 58–71) and actinide (elements 90–103) series are given.

The Periodic Table is used throughout the chemistry course and provides detailed information about individual elements and periodic trends.

of the Elements

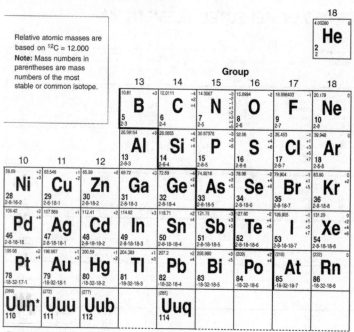

*The systematic names and symbols for elements of atomic numbers above 109 will be used until the approval of trivial names by IUPAC.

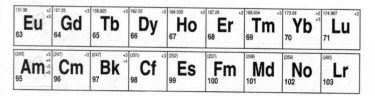

Table S—Properties of Selected Elements

Table S provides a listing of elements whose atomic numbers range from 1 through 57, and from 72 through 89. The listing is in order of increasing atomic number, and it provides the symbol and name of each element, as well as the following information:

- The *ionization energy* (in kJ/mol), which can be used to compare the ease with which a series of atoms lose their first electrons.
- The *electronegativity*, which can be used to predict the ionic character of a pair of bonded atoms.

PROPERTIES OF SELECTED ELEMENTS

Atomic Number	Symbol	Name	Ionization Energy (kJ/mol)	Electro-negativity	Melting Point (K)	Boiling Point (K)	Density** (g/cm³)	Atomic Radius (pm)
1	H	hydrogen	1312	2.1	14	20	0.00009	208
2	He	helium	2372	—	1	4	0.000179	50
3	Li	lithium	520	1.0	454	1620	0.534	155
4	Be	beryllium	900	1.6	1551	3243	1.8477	112
5	B	boron	801	2.0	2573	3931	2.340	98
6	C	carbon	1086	2.6	3820	5100	3.513	91
7	N	nitrogen	1402	3.0	63	77	0.00125	92
8	O	oxygen	1314	3.4	55	90	1.429	65
9	F	fluorine	1681	4.0	54	85	0.001696	57
10	Ne	neon	2081	—	24	27	0.0009	51
11	Na	sodium	496	0.9	371	1156	0.971	190
12	Mg	magnesium	736	1.3	922	1363	1.738	160
13	Al	aluminum	578	1.6	934	2740	2.698	143
14	Si	silicon	787	1.9	1683	2628	2.329	132
15	P	phosphorus	1012	2.2	44	553	1.820	128
16	S	sulfur	1000	2.6	386	718	2.070	127
17	Cl	chlorine	1251	3.2	172	239	0.003214	97
18	Ar	argon	1521	—	84	87	0.001783	88
19	K	potassium	419	0.8	337	1047	0.862	235
20	Ca	calcium	590	1.0	1112	1757	1.550	197
21	Sc	scandium	633	1.4	1814	3104	2.989	162
22	Ti	titanium	659	1.5	1933	3580	4.540	145
23	V	vanadium	651	1.6	2160	3650	6.100	134
24	Cr	chromium	653	1.7	2130	2945	7.190	130
25	Mn	manganese	717	1.6	1517	2235	7.440	135
26	Fe	iron	762	1.8	1808	3023	7.874	126
27	Co	cobalt	760	1.9	1768	3143	8.900	125
28	Ni	nickel	737	1.9	1726	3005	8.902	124
29	Cu	copper	745	1.9	1357	2840	8.960	128
30	Zn	zinc	906	1.7	693	1180	7.133	138
31	Ga	gallium	579	1.8	303	2676	5.907	141
32	Ge	germanium	762	2.0	1211	3103	5.323	137
33	As	arsenic	944	2.2	1090	889	5.780	139
34	Se	selenium	941	2.6	490	958	4.790	140
35	Br	bromine	1140	3.0	266	332	3.122	112
36	Kr	krypton	1351	—	117	121	0.00375	103
37	Rb	rubidium	403	0.8	312	961	1.532	248
38	Sr	strontium	549	1.0	1042	1657	2.540	215
39	Y	yttrium	600	1.2	1795	3611	4.469	178
40	Zr	zirconium	640	1.3	2125	4650	6.506	160

- The *melting point* and *boiling point* (in K) at standard pressure.
- The *density* (in g/cm3) at standard temperature and pressure (STP), which can be used to relate the mass of a sample of an element to its volume.
- The *atomic radius* (in pm), which can be used to compare the sizes of individual atoms or examine periodic trends in atomic size down a Group and across a Period.

PROPERTIES OF SELECTED ELEMENTS (continued)

Atomic Number	Symbol	Name	Ionization Energy (kJ/mol)	Electro-negativity	Melting Point (K)	Boiling Point (K)	Density** (g/cm^3)	Atomic Radius (pm)
41	Nb	niobium	652	1.6	2741	5015	8.570	146
42	Mo	molybdenum	684	2.2	2890	4885	10.220	139
43	Tc	technetium	702	1.9	2445	5150	11.500	136
44	Ru	ruthenium	710	2.2	2583	4173	12.370	134
45	Rh	rhodium	720	2.3	2239	4000	12.410	134
46	Pd	palladium	804	2.2	1825	3413	12.020	137
47	Ag	silver	731	1.9	1235	2485	10.500	144
48	Cd	cadmium	868	1.7	594	1038	8.650	171
49	In	indium	558	1.8	429	2353	7.310	166
50	Sn	tin	709	2.0	505	2543	7.310	162
51	Sb	antimony	831	2.1	904	1908	6.691	159
52	Te	tellurium	869	2.1	723	1263	6.240	142
53	I	iodine	1008	2.7	387	458	4.930	132
54	Xe	xenon	1170	2.6	161	166	0.0059	124
55	Cs	cesium	376	0.8	302	952	1.873	267
56	Ba	barium	503	0.9	1002	1910	3.594	222
57	La	lanthanum	538	1.1	1194	3730	6.145	138
Elements 58–71 have been omitted.								
72	Hf	hafnium	659	1.3	2503	5470	13.310	167
73	Ta	tantalum	728	1.5	3269	5698	16.654	149
74	W	tungsten	759	2.4	3680	5930	19.300	141
75	Re	rhenium	756	1.9	3453	5900	21.020	137
76	Os	osmium	814	2.2	3327	5300	22.590	135
77	Ir	iridium	865	2.2	2683	4403	22.560	136
78	Pt	platinum	864	2.3	2045	4100	21.450	139
79	Au	gold	890	2.5	1338	3080	19.320	146
80	Hg	mercury	1007	2.0	234	630	13.546	160
81	Tl	thallium	589	2.0	577	1730	11.850	171
82	Pb	lead	716	2.3	601	2013	11.350	175
83	Bi	bismuth	703	2.0	545	1833	9.747	170
84	Po	polonium	812	2.0	527	1235	9.320	167
85	At	astatine	—	2.2	575	610	—	145
86	Rn	radon	1037	—	202	211	0.00973	134
87	Fr	francium	393	0.7	300	950	—	270
88	Ra	radium	—	0.9	973	1413	5.000	233
89	Ac	actinium	499	1.1	1320	3470	10.060	—
Elements 90 and above have been omitted.								

*Boiling point at standard pressure
**Density at STP

Table T—Important Formulas and Equations

Table T provides a listing of the important formulas and equations that are part of the chemistry course. The next section explains how these equations might be used in solving problems.

IMPORTANT FORMULAS AND EQUATIONS

1. Density	$d = \dfrac{m}{V}$	d = density m = mass V = volume
2. Mole Calculations	number of moles = $\dfrac{\text{given mass (g)}}{\text{gram-formula mass}}$	
3. Percent Error	% error = $\dfrac{\text{measured value} - \text{accepted value}}{\text{accepted value}} \times 100$	
4. Percent Composition	% composition by mass = $\dfrac{\text{mass of part}}{\text{mass of whole}} \times 100$	
5. Concentration	parts per million = $\dfrac{\text{grams of solute}}{\text{grams of solution}} \times 1{,}000{,}000$ molarity = $\dfrac{\text{moles of solute}}{\text{liters of solution}}$	
6. Combined Gas Law	$\dfrac{P_1V_1}{T_1} = \dfrac{P_2V_2}{T_2}$	P = pressure V = volume T = temperature (K)
7. Titration	$M_AV_A = M_BV_B$	M_A = molarity of H^+ M_B = molarity of OH^- V_A = volume of acid V_B = volume of base
8. Heat	$q = mC\Delta T$ $q = mH_f$ $q = mH_v$	q = heat H_f = heat of fusion m = mass H_v = heat of vaporization C = specific heat capacity ΔT = change in temperature
9. Temperature	$K = {}^{\circ}C + 273$	K = kelvin $^{\circ}C$ = degrees Celsius
10. Radioactive Decay	fraction remaining = $\left(\dfrac{1}{2}\right)^{\frac{t}{T}}$ number of half-life periods = $\dfrac{t}{T}$	t = total time elapsed T = half-life

Using the Equations to Solve Chemistry Problems

IMPORTANT FORMULAS/EQUATIONS

In this section, various problems are solved using the formulas and equations found in Reference Table T.

1. Density: $d = \dfrac{m}{v}$

Problem

Calculate the mass of a sample of the element niobium (Atomic Number = 41) if the volume of the sample is 13.00 cm³.

Solution

- Use Reference Table S to find the density of niobium (8.570 g/cm³).
- Rearrange the variables in the equation in order to solve for mass: $m = d \cdot v$.
- Substitute the data in the equation and perform the calculation:

$$m = (8.570 \ \frac{g}{cm^3}) \cdot (13.00 \ cm^3) = \textbf{111.4 g}$$

----

2. Mole Calculations: $\text{number of moles} = \dfrac{\text{given mass}}{\text{gram-formula mass}}$

Problem

Calculate the number of moles in a sample of oxygen gas (O_2, gram-formula mass = 32.0 g/mol) if its mass is 56.0 grams.

Solution
- Substitute the data into the equation:

$$\text{number of moles} = \frac{56.0 \text{ g}}{32.0 \frac{\text{g}}{\text{mol}}} = 1.75 \text{ mol}$$

---

3. Percent Error: $\% \text{ error} = \dfrac{\text{measured value} - \text{accepted value}}{\text{accepted value}} \times 100$

Problem
A student measured the density of a sample of iron and obtained a value of 8.391 g/cm³. What is the percent error of the student's determination?

Solution
- Use Reference Table S to obtain the density of iron (7.874 g/cm³).
- Substitute the data into the equation:

$$\% \text{ error} = \frac{8.391 \frac{\text{g}}{\text{cm}^3} - 7.874 \frac{\text{g}}{\text{cm}^3}}{7.874 \frac{\text{g}}{\text{cm}^3}} \times 100 = \mathbf{6.566\%}$$

---

4. Percent Composition: $\% \text{ composition by mass} =$
$$\frac{\text{mass of part}}{\text{mass of whole}} \times 100$$

Problem
Calculate the percent composition by mass of carbon in $C_6H_{12}O_6$.

Solution
- Use the Periodic Table to determine the mass of one mole of the compound (180.2 g) and the mass of the *carbon* in 1 mole of the compound (72.07 g).
- Substitute the data into the equation:

$$\% \text{ composition} = \frac{72.07 \text{ g}}{180.2 \text{ g}} \times 100 = \mathbf{39.99\%}$$

5a. Parts per Million: $\dfrac{\text{grams of solute}}{\text{grams of solution}} \times 1{,}000{,}000$

Problem

The concentration of arsenic in a certain river is 0.267 gram per 2,000 grams of river water. What is the arsenic concentration in parts per million (ppm)?

Solution

• Substitute the data into the equation:

$$\frac{0.267 \text{ g}}{2{,}000 \text{ g}} \times 1{,}000{,}000 = \textbf{134 ppm}$$

5b. Molarity: $\text{molarity} = \dfrac{\text{grams of solute}}{\text{liters of solution}}$

Problem

What is the molarity of an aqueous solution of ammonia (gram-formula mass = 17.03 g/mol) if 30.8 grams of solute are dissolved in 2.50 liters of solution?

Solution

• Use the gram-formula mass to convert the mass to moles.
• Substitute the data into the equation:

$$\text{molarity} = \frac{(30.8 \text{ g}) \cdot \left(\dfrac{1 \text{ mol}}{17.03 \text{ g}}\right)}{} = \textbf{0.723 M}$$

----

6. Combined Gas Law: $\dfrac{P_1 V_1}{T_1} = \dfrac{P_2 V_2}{T_2}$

Problem

A 50.0-milliliter sample, initially at STP, has its pressure changed to 0.85 atmosphere and its temperature changed to 330 K. What is the new volume of the gas?

Solution
- Use Reference Table *A* to obtain the STP values for temperature (273 K) and pressure (1 atm).
- Rearrange the equation in order to solve for V_2: $V_2 = \dfrac{P_1 V_1 T_2}{T_1 P_2}$.
- Substitute the data into the equation:

$$V_2 = \frac{(1 \text{ atm}) \cdot (50.0 \text{ mL}) \cdot (330. \text{ K})}{(273 \text{ K}) \cdot (0.85 \text{ atm})} = \mathbf{71.1 \text{ mL}}$$

---

7. Titration (Acid–Base): $M_A V_A = M_B V_B$

Problem
How many milliliters of 0.25-molar NaOH are needed to neutralize 75 milliliters of 0.17-molar HCl?

Solution
- Rearrange the equation to solve for V_B: $V_B = \dfrac{M_A V_A}{M_B}$.
- Substitute the data into the equation:

$$V_B = \frac{(0.17 \text{ M}) \cdot (75 \text{ mL})}{(0.25 \text{ M})} = \mathbf{51 \text{ mL (NaOH)}}$$

8a. Heat Transferred: $q = mC\Delta T$

Problem
How much heat is released by 120 grams of liquid water as it cools from 340 K to 290 K?

Solution
- Use Reference Table *B* to obtain the specific heat capacity of liquid water (4.2 J/g·K).
- Substitute the data into the equation:

$$q = (120 \text{ g}) \cdot \left(4.2 \, \frac{\text{J}}{\text{g} \cdot \text{K}} \right) \cdot (290 \text{ K} - 340 \text{ K}) = \mathbf{-25{,}000 \text{ J}}$$

(The minus sign indicates that heat is *released*.)

8b. Heat of Fusion: $q = mH_f$

Problem

How much heat is absorbed by 200.0 grams of ice as it melts at 273 K?

Solution

- Use Reference Table B to obtain the value for the heat of fusion of ice (333.6 J/g).
- Substitute the data into the equation:

$$q = (200.0 \text{ g}) \cdot \left(333.6 \frac{\text{J}}{\text{g}}\right) = \mathbf{66{,}720 \text{ J}}$$

8c. Heat of Vaporization: $q = mH_v$

Problem

How many grams of steam at 100°C will be condensed to liquid water if 4,000. joules of heat are released?

Solution

- Use Reference Table B to obtain the value for the heat of vaporization of water (2,259 J/g).
- Rearrange the equation to solve for mass: $m = \dfrac{q}{H_v}$
- Substitute the data into the equation:

$$m = \frac{4{,}000. \text{ J}}{\left(2259 \frac{\text{J}}{\text{g}}\right)} = \mathbf{1.771 \text{ g}}$$

----

9. Temperature: K = °C + 273

Problem

What is the Kelvin temperature equivalent of 34°C?

Solution

- Substitute the data into the equation:

K = 34°C + 273 = **307 K**

10a. Radioactive Decay: fraction remaining = $\left(\dfrac{1}{2}\right)^{\frac{t}{T}}$

10b. Radioactive Decay: number of half-life periods = $\dfrac{t}{T}$

Problem
What fraction of a sample of ^{42}K will remain unchanged after 37.2 hours of decay?

Solution
- Use Reference Table N to obtain the half-life of ^{42}K (12.4 h).
- Use equation 10b to determine the number of half-life periods:

    number of half-life periods = $\dfrac{37.2 \text{ h}}{12.4 \text{ h}}$ = 3.00 = **3**

- Substitute the number of half-life periods into equation 10a:

    fraction remaining = $\left(\dfrac{1}{2}\right)^{3}$ = $\dfrac{1}{8}$

That is, one-eighth of the original sample of ^{42}K remains unchanged after 37.2 hours of decay.

Regents Examinations, Answers and Self-Analysis Charts

Examination August 2002

Chemistry
The Physical Setting

PART A

Answer all questions in this part.

Directions (1–30): For *each* statement of question, write in the answer space the *number* of the word or expression that, of those given, best completes the statement or answers the question. Some questions may require the use of the *Reference Tables for Physical Setting/Chemistry*.

1 Subatomic particles can usually pass undeflected through an atom because the volume of an atom is composed of

(1) an uncharged nucleus
(2) largely empty space
(3) neutrons
(4) protons

1 _____

2 What is the total number of electrons in the valence shell of an atom of aluminum in the ground state?

(1) 8   (3) 3
(2) 2   (4) 10

2 _____

3 Which of these elements has physical and chemical properties most similar to silicon (Si)?

    (1) germanium (Ge)
    (2) lead (Pb)
    (3) phosphorus (P)
    (4) chlorine (Cl)   3 _____

4 What is the total number of protons in the nucleus of an atom of potassium-42?

    (1) 15   (3) 39
    (2) 19   (4) 42   4 _____

5 Given the equation: $H_2O(s) \rightleftharpoons H_2O(\ell)$

At which temperature will equilibrium exist when the atmospheric pressure is 1 atm?

    (1) 0 K   (3) 273 K
    (2) 100 K   (4) 373 K   5 _____

6 Which species represents a chemical compound?

    (1) N_2 (3) Na
    (2) NH_4^+ (4) $NaHCO_3$ 6 _____

7 Which mixture can be separated by using the equipment shown below?

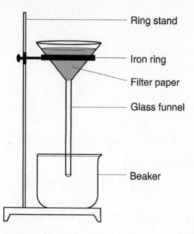

(1) NaCl(aq) and SiO_2(s)
(2) NaCl(aq) and $C_6H_{12}O_6$(aq)
(3) CO_2(aq) and NaCl(aq)
(4) CO_2(aq) and $C_6H_{12}O_6$(aq)

7 _____

8 Which reaction represents natural nuclear decay?

(1) $H^+ + OH^- \rightarrow H_2O$
(2) $KClO_3 \rightarrow K^+ + ClO_3^-$
(3) $^{235}_{92}U \rightarrow ^{4}_{2}He + ^{231}_{90}Th$
(4) $^{14}_{7}N + ^{4}_{2}He \rightarrow ^{17}_{8}O + ^{1}_{1}H$

8 _____

9 If an equation is balanced properly, both sides of the equation must have the same number of

(1) atoms
(2) coefficients
(3) molecules
(4) moles of molecules

9 _____

10 Which of the following elements has the highest electronegativity?

(1) H   (3) Al
(2) K   (4) Ca

10 _____

11 Which formula represents an ionic compound?

    (1) NaCl   (3) HCl

    (2) N_2O (4) H_2O 11 _____

12 Which species does *not* have a noble gas electron configuration?

    (1) Na^+ (3) Ar

    (2) Mg^{2+} (4) S   12 _____

13 Which statement correctly describes a chemical reaction at equilibrium?

    (1) The concentrations of the products and reactants are equal.

    (2) The concentrations of the products and reactions are constant.

    (3) The rate of the forward reaction is less than the rate of the reverse reaction.

    (4) The rate of the forward reaction is greater than the rate of the reverse reaction.   13 _____

14 Given the reaction:

$$CH_4(g) + 2\ O_2(g) \rightarrow 2\ H_2O(g) + CO_2(g)$$

What is the overall result when $CH_4(g)$ burns according to this reaction?

    (1) Energy is absorbed and ΔH is negative.

    (2) Energy is absorbed and ΔH is positive.

    (3) Energy is released and ΔH is negative.

    (4) Energy is released and ΔH is positive.   14 _____

15 A hydrated salt is a solid that includes water molecules within its crystal structure. A student heated a 9.10-gram sample of a hydrated salt to a constant mass of 5.41 grams. What percent by mass of water did the salt contain?

(1) 3.69%   (3) 40.5%

(2) 16.8%   (4) 59.5%   15 _____

16 Which statement correctly describes a sample of gas confined in a sealed container?

(1) It always has a definite volume, and it takes the shape of the container.

(2) It takes the shape and the volume of any container in which it is confined.

(3) It has a crystalline structure.

(4) It consists of particles arranged in a regular geometric pattern.   16 _____

17 Which molecule contains a triple covalent bond?

(1) H_2 (3) O_2

(2) N_2 (4) Cl_2 17 _____

18 The solid and liquid phases of water can exist in a state of equilibrium at 1 atmosphere of pressure and a temperature of

(1) 0°C   (3) 273°C

(2) 100°C   (4) 373°C   18 _____

19 Which compound is an alcohol?

(1) propanal   (3) butane

(2) ethyne   (4) methanol   19 _____

20 In which reaction is soap a product?

(1) addition   (3) saponification

(2) substitution   (4) polymerization   20 _____

21 The spontaneous decay of an atom is called

(1) ionization
(3) combustion
(2) crystallization
(4) transmutation   21 _____

22 In any redox reaction, the substance that undergoes reduction will

(1) lose electrons and have a decrease in oxidation number
(2) lose electrons and have an increase in oxidation number
(3) gain electrons and have a decrease in oxidation number
(4) gain electrons and have an increase in oxidation number   22 _____

23 Which electron configuration is correct for a sodium ion?

(1) 2–7
(3) 2–8–1
(2) 2–8
(4) 2–8–2   23 _____

24 In which equation does the term "heat" represent heat of fusion?

(1) $NaCl(s) + heat \rightarrow NaCl(\ell)$
(2) $NaOH(aq) + HCl(aq) \rightarrow NaCl(aq) + H_2O(\ell) + heat$
(3) $H_2O(\ell) + heat \rightarrow H_2O(g)$
(4) $H_2O(\ell) + HCl(g) \rightarrow H_3O^+(aq) + Cl^-(aq) + heat$ 24 _____

25 Which substance is an Arrhenius acid?

(1) $LiF(aq)$
(3) $Mg(OH)_2(aq)$
(2) $HBr(aq)$
(4) CH_3CHO 25 _____

26 Which type of emission has the highest penetrating power?

(1) alpha
(3) positron
(2) beta
(4) gamma   26 _____

Note that questions 27 through 30 have only three choices.

27 As the elements in Group 17 are considered in order of increasing atomic number, the chemical reactivity of each successive element

(1) decreases
(2) increases
(3) remains the same

27 _____

28 As the pressure on the surface of a liquid *decreases*, the temperature at which the liquid will boil

(1) decreases
(2) increases
(3) remains the same

28 _____

29 As a Ca atom undergoes oxidation to Ca^{2+}, the number of neutrons in its nucleus

(1) decreases
(2) increases
(3) remains the same

29 _____

30 As the temperature of a liquid increases, its vapor pressure

(1) decreases
(2) increases
(3) remains the same

30 _____

PART B–1

Answer all questions in this part.

Directions (31–50): For *each* statement or question, write in the answer space the *number* of the word or expression that, of those given, best completes the statement or answers the question. Some questions may require the use of the *Reference Tables for Physical Setting/Chemistry*.

31 Compared to the nonmetals in Period 2, the metals in Period 2 generally have larger

    (1) ionization energies
    (2) electronegativities
    (3) atomic radii
    (4) atomic numbers   31 _____

32 Which of the following Group 2 elements has the *lowest* first ionization energy?

    (1) Be   (3) Ca
    (2) Mg   (4) Ba   32 _____

33 The table below shows the normal boiling point of four compounds.

Compound	Normal Boiling Point (°C)
$HF(\ell)$	19.4
$CH_3Cl(\ell)$	−24.2
$CH_3F(\ell)$	−78.6
$HCl(\ell)$	−83.7

Which compound has the strongest intermolecular forces?

    (1) $HF(\ell)$ (3) $CH_3F(\ell)$
    (2) $CH_3Cl(\ell)$ (4) $HCl(\ell)$ 33 _____

34 According to Table I, which salt releases energy as it dissolves?

(1) KNO_3 (3) NH_4NO_3

(2) LiBr   (4) NaCl   34 _____

35 During a laboratory activity, a student combined two solutions. In the laboratory report, the student wrote "A yellow color appeared." The statement represents the student's recorded

(1) conclusion   (3) hypothesis

(2) observation   (4) inference   35 _____

36 How many moles of solute are contained in 200 milliliters of a 1 M solution?

(1) 1   (3) 0.8

(2) 0.2   (4) 200   36 _____

37 Increasing the temperature increases the rate of a reaction by

(1) lowering the activation energy

(2) increasing the activation energy

(3) lowering the frequency of effective collisions between reacting molecules

(4) increasing the frequency of effective collisions between reacting molecules   37 _____

38 Given the equilibrium reaction in a closed system:

$$H_2(g) + I_2(g) + heat \rightleftharpoons 2HI(g)$$

What will be the result of an increase in temperature?

(1) The equilibrium will shift to the left and $[H_2]$ will increase.

(2) The equilibrium will shift to the left and $[H_2]$ will decrease.

(3) The equilibrium will shift to the right and $[HI]$ will increase.

(4) The equilibrium will shift to the right and $[HI]$ will decrease.   38 _____

39 Which sample has the *lowest* entropy?

    (1) 1 mole of KNO₃(ℓ)   (3) 1 mole of H₂O(ℓ)

    (2) 1 mole of KNO₃(s)   (4) 1 mole of H₂O(g)   39 _____

40 Which of the following compounds is *least* soluble in water?

    (1) copper (II) chloride

    (2) aluminum acetate

    (3) iron (III) hydroxide

    (4) potassium sulfate   40 _____

41 According to Table I, which potential energy diagram best represents the reaction that forms $H_2O(\ell)$ from its elements?

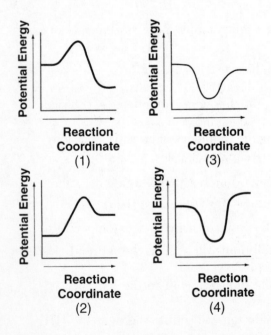

41 _____

42 Which structural formula is *incorrect*?

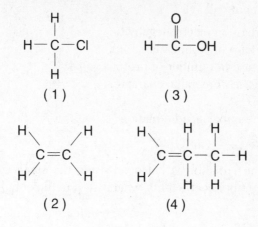

(1)   (3)

(2)   (4)

42 _____

43 Given the fusion reaction:

$$^2_1H + ^2_1H \rightarrow X + energy$$

Which particle is represented by X?

(1) ^1_1H (3) ^3_2He
(2) ^3_1H (4) ^4_2He 43 _____

44 The vapor pressure of a liquid is 0.92 atm at 60°C. The normal boiling point of the liquid could be

(1) 35°C   (3) 55°C
(2) 45°C   (4) 65°C   44 _____

45 When 50. milliliters of an HNO_3 solution is exactly neutralized by 150 milliliters of a 0.50 M solution of KOH, what is the concentration of HNO_3?

(1) 1.0 M   (3) 3.0 M
(2) 1.5 M   (4) 0.5 M   45 _____

46 In Period 3, from left to right in order, each successive element will

  (1) decrease in electronegativity
  (2) decrease in atomic mass
  (3) increase in number of protons
  (4) increase in metallic character   46 _____

47 Given the unbalanced equation:

$$\_\_ Al + \_\_ CuSO_4 \rightarrow \_\_ Al_2(SO_4)_3 + \_\_ Cu$$

When the equation is balanced using the *smallest* whole-number coefficients, what is the coefficient of Al?

  (1) 1   (3) 3
  (2) 2   (4) 4   47 _____

48 One hundred grams of water is saturated with NH_4Cl at 50°C. According to Table G, if the temperature is lowered to 10°C, what is the total amount of NH_4Cl that will precipitate?

  (1) 5.0 g   (3) 30. g
  (2) 17 g   (4) 50. g   48 _____

49 What is the total number of grams of NaI(s) needed to make 1.0 liter of a 0.010 M solution?

  (1) 0.015   (3) 1.5
  (2) 0.15   (4) 15   49 _____

50 Given the reaction:

$$2H_2(g) + O_2(g) \rightarrow 2H_2O(\ell) + 571.6 \text{ kJ}$$

What is the approximate ΔH for the formation of 1 mole of $H_2O(\ell)$?

  (1) −285.8 kJ   (3) −571.6 kJ
  (2) +285.8 kJ   (4) +571.6 kJ   50 _____

PART B-2

Answer all questions in this part.

Directions (51–57): Record your answers on the answer sheet provided in the back. Some questions may require the use of the *Reference Tables for Physical Setting/Chemistry*.

51 On a field trip, Student *X* and Student *Y* collected two rock samples. Analysis revealed that both rocks contained lead and sulfur. One rock contained a certain percentage of lead and sulfur by mass, and the other rock contained a different percentage of lead and sulfur by mass. Student *X* stated that the rocks contained two different mixtures of lead and sulfur. Student *Y* stated that the rocks contained two different compounds of lead and sulfur. Their teacher stated that both students could be correct.

Draw particle diagrams in *each* of the rock diagrams provided *in your answer booklet* to show how Student *X*'s and Student *Y*'s explanations could both be correct. Use the symbols in the key provided *in your answer booklet* to sketch lead and sulfur atoms. [2]

52 One electron is removed from both an Na atom and a K atom, producing two ions. Using principles of atomic structure, explain why the Na ion is much smaller than the K ion. Discuss both ions in your answer. [2]

53 In the space provided *in your answer booklet*, draw an electron-dot diagram for *each* of the following substances:

a calcium oxide (an ionic compound) [1]
b hydrogen bromide [1]
c carbon dioxide [1]

54 A sample of water is heated from a liquid at 40°C to a gas at 110°C. The graph of the heating curve is shown in your answer booklet.

   a On the heating curve diagram provided *in your answer booklet*, label *each* of the following regions: [1]
       Liquid, only
       Gas, only
       Phase change

   b For section QR of the graph, state what is happening to the water molecules as heat is added. [1]

   c For section RS of the graph, state what is happening to the water molecules as heat is added. [1]

55 Given the structural formula for butane:

$$
\begin{array}{c}
\ \ \ \text{H} \ \ \ \text{H} \ \ \ \text{H} \ \ \ \text{H} \\
\ \ \ | \ \ \ \ | \ \ \ \ | \ \ \ \ | \\
\text{H}-\text{C}-\text{C}-\text{C}-\text{C}-\text{H} \\
\ \ \ | \ \ \ \ | \ \ \ \ | \ \ \ \ | \\
\ \ \ \text{H} \ \ \ \text{H} \ \ \ \text{H} \ \ \ \text{H}
\end{array}
$$

In the space provided *in your answer booklet*, draw the structural formula of an isomer of butane. [1]

56 Given the ester: ethyl butanoate

   a In the space provided *in your answer booklet*, draw the structural formula for this ester. [1]

   b Determine the gram formula mass of this ester. [1]

57 Given the reaction: $4 \text{ Al(s)} + 3 \text{ O}_2\text{(g)} \rightarrow 2 \text{ Al}_2\text{O}_3\text{(s)}$

   a Write the balanced oxidation half-reaction for this oxidation-reduction reaction. [1]

   b What is the oxidation number of oxygen in Al_2O_3? [1]

PART C

Answer all questions in this part.

Directions (58–62): Record your answers on the answer sheet provided in the back. Some questions may require the use of the *Reference Tables for Physical Setting/Chemistry*.

58 *a* State one possible advantage of using nuclear power instead of burning fossil fuels. [1]

   b State one possible risk of using nuclear power. [1]

   c If animals feed on plants that have taken up Sr-90, the Sr-90 can find its way into their bone structure. Explain one danger to the animals. [1]

59 Four flasks each contain 100 milliliters of aqueous solutions of equal concentrations at 25°C and 1 atm.

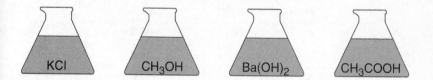

a Which solutions contain electrolytes? [1]

b Which solution has the *lowest* pH? [1]

c What causes some aqueous solutions to have a low pH? [1]

d Which solution is most likely to react with an Arrhenius acid to form a salt and water? [1]

e Which solution has the *lowest* freezing point? Explain your answer. [2]

60 The equation for the saturated solution equilibrium of potassium nitrate (KNO_3) is shown below.

$$KNO_3(s) + energy \overset{H_2O}{\rightleftharpoons} K^+(aq) + NO_3^-(aq)$$

a In the space provided *in your answer booklet*, diagram the products. Use the key provided *in your answer booklet*. Indicate the exact arrangement of the particles you diagram.   [2]

b Compare the rate of dissolving KNO_3 with the rate of recrystallization of KNO_3 for the saturated solution.   [1]

61 Electron affinity is defined as the energy released when an atom and an electron react to form a negative ion. The data for Group 1 elements are presented below.

Element	Atomic Number	Electron Affinity in kJ/mole
Cs	55	45.5
H	1	72.8
K	19	46.4
Li	3	59.8
Na	11	52.9
Rb	37	?

On the grid provided *in your answer booklet*, draw a graph to show the relationship between *each* member of Group 1 and its electron affinity by following the directions below.

a Label the *y*-axis "Electron Affinity" and choose an appropriate scale. Label the *x*-axis "Atomic Number" and choose an appropriate scale.   [1]

b Plot the data from the data table and connect the points with straight lines.   [1]

c Using your graph, estimate the electron affinity of Rb, in kilojoules/mole.   [1]

62 A student used a balance and a graduated cylinder to collect the following data:

Sample mass	10.23 g
Volume of water	20.0 mL
Volume of water and sample	21.5 mL

a Calculate the density of the element. Show your work. Include the appropriate number of significant figures and proper units.   [3]

b If the accepted value is 6.93 grams per milliliter, calculate the percent error.   [1]

c What error is introduced if the volume of the sample is determined first?   [1]

Answer Sheet
August 2002

Chemistry
The Physical Setting

PART B–2

Answer all questions in Part B–2 and Part C. Record your answers on the answer sheet.

51

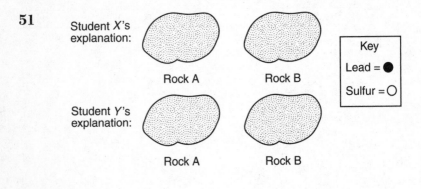

Student *X*'s explanation:

Rock A   Rock B

Key
Lead = ●
Sulfur = ○

Student *Y*'s explanation:

Rock A   Rock B

52 _____

_____

53 a   b   c

54 a

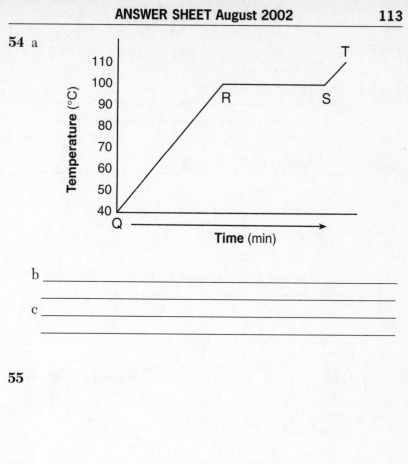

b _____

_____

c _____

_____

55

56 a

b _____

57 a _____

b _____

58 a _____

    b _____

    c _____

59 a _____
    b _____
    c _____

    d _____
    e _____

60 a

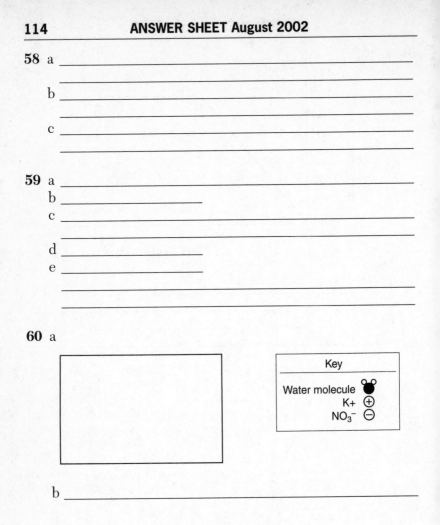

Key	
Water molecule	
K+	⊕
NO₃⁻	⊖

    b _____

61 a and b

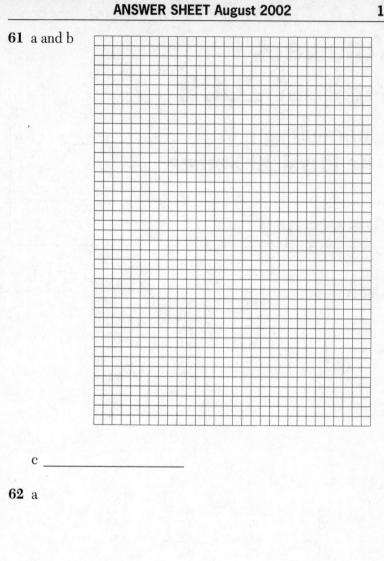

c _____

62 a

b _____
c _____

Answers
August 2002

Chemistry
The Physical Setting

Answer Key

PART A

1. 2	7. 1	13. 2	19. 4	25. 2
2. 3	8. 3	14. 3	20. 3	26. 4
3. 1	9. 1	15. 3	21. 4	27. 1
4. 2	10. 1	16. 2	22. 3	28. 1
5. 3	11. 1	17. 2	23. 2	29. 3
6. 4	12. 4	18. 1	24. 1	30. 2

PART B–1

31. 3	35. 2	39. 2	43. 4	47. 2
32. 4	36. 2	40. 3	44. 4	48. 2
33. 1	37. 4	41. 1	45. 2	49. 3
34. 2	38. 3	42. 4	46. 3	50. 1

Answers Explained

PART A

1. **2** The modern nuclear model of the atom consists of a small, positively charged nucleus surrounded by mostly empty space. The electrons are located in the space outside the nucleus.

2. **3** The valence shell is the outermost shell of electrons. The ground state electron configuration of aluminum (Al) is given in the Periodic Table of the Elements: 2–8–3. Therefore, the valence shell contains 3 electrons.

3. **1** Elements that are located within a single group have similar physical and chemical properties. In addition, elements that are separated by only one period have the most similar properties. Silicon and germanium are both located in Group 14 and are in Periods 3 and 4, respectively.

Wrong Choice Explained:
(2) Although lead is also located in Group 14, it is located in Period 6. Therefore, its properties are less similar to the properties of silicon.

4. **2** The designation potassium-42 represents an isotope of potassium (K) that contains a total of 42 protons and neutrons in its nucleus. The number of protons in the nucleus is determined by the atomic number of the element. According to the Periodic Table of the Elements, potassium has an atomic number of 19.

5. **3** The equation represents the phase equilibrium between ice and liquid water. At 1 atmosphere, this equilibrium occurs at a temperature of 273 K (or 0°C).

Wrong Choices Explained:
(1) 0 K is known as *absolute zero* and is the lowest limit of temperature.
(4) 373 K (or 100°C) is the temperature at which liquid water and steam are in equilibrium at 1 atmospheric pressure.

6. **4** A chemical compound is a combination of elements whose composition is fixed, is electrically neutral, and can be represented by a formula. Only choice (4), $NaHCO_3$, meets these three criteria.

Wrong Choices Explained:
(1), (3) N_2 and Na are *elements*.
(2) NH_4^+ is a polyatomic *ion*.

7. **1** The filtration equipment shown in the diagram is used to separate undissolved solids from substances that are dissolved in solution. Of the choices given, only choice (1), NaCl(aq) and SiO_2(s), contains an undissolved solid.

8. **3** Natural nuclear decay involves the spontaneous change of a radioactive atom without the presence of a bombarding particle, as shown in choice (3).

Wrong Choices Explained:
(1) This equation represents neutralization; it is not a nuclear reaction.
(2) This equation represents the dissociation of $KClO_3$ into ions; it is not a nuclear reaction.
(4) This nuclear equation represents the transmutation of nitrogen-14 by an alpha particle; it does not represent natural decay.

9. **1** Properly balanced equations conserve mass, electric charge, and energy. As a result, the numbers of atoms on both sides of the balanced equation must be equal.

Wrong Choices Explained:
(2), (3), (4) Since molecules differ in composition, the number of molecules, the total moles of molecules, and the coefficients need not be conserved in a chemical equation. As an example, consider the equation $2O_3 \rightarrow 3O_2$. Each side of the equation contains 6 moles of oxygen *atoms*. However, the numbers of molecules—and their coefficients—are not equal.

10. **1** Use Reference Table S. The element H has the highest electronegativity (2.1). You could have arrived at the answer without using the table. Metals have low electronegativities; only choice (1), H, is a *nonmetal*.

11. **1** An ionic compound consists of a metal and a nonmetal with a significant difference in electronegativity (≥ 1.7). Of the choices given, only choice (1), NaCl, meets these criteria.

12. **4** Use the Periodic Table of the Elements. The element S has the configuration 2–8–6, which is not a noble gas (Group 18) configuration.

Wrong Choices Explained:

(1) An Na^+ ion is formed by removing one electron from Na; its electron configuration is $(2-8-1) - 1 = 2-8$, which is the electron configuration of Ne, a noble gas.

(2) An Mg^{2+} ion is formed by removing two electrons from Mg; its electron configuration is $(2-8-2) - 2 = 2-8$, which is the electron configuration of Ne, a noble gas.

(3) Ar is a noble gas.

13. **2** When chemical equilibrium is reached, the rates of the forward and reverse reactions are equal. As a result, the concentrations of the reactants and products are constant.

14. **3** Refer to the first reaction in Reference Table *I*. This reaction is very similar to the one given in this question. In general, combustion (burning) reactions release a great deal of energy and have large negative heats of reaction (ΔH).

15. **3** The difference between the masses of the hydrated and anhydrous crystals is the mass of the water that is present in the hydrated crystals.

$$\text{Mass of water} = 9.10 \text{ g} - 5.41 \text{ g} = 3.69 \text{ g}$$

The percent of water present in the hydrated salt is calculated as follows:

$$\text{percent water} = \frac{\text{mass of water}}{\text{mass of hydrated crystals}} \times 100$$

$$= \frac{3.69 \text{ g}}{9.10 \text{ g}} \times 100$$

$$= \mathbf{40.5\%}$$

16. **2** Gases are classified as samples of matter that take the shape and volume of the container that encloses them.

17. **2** Refer to the Periodic Table of the Elements. An atom of nitrogen has five valence electrons, and three of them are unpaired. In forming the N_2 molecule, three pairs of electrons are shared by both atoms. The Lewis electron-dot structure of N_2 is shown below:

$$\overset{\bullet\bullet}{N} \equiv \overset{\bullet\bullet}{N}$$

Wrong Choices Explained:

(1), (4) The atoms in H_2 and Cl_2 are joined by *single* bonds.

(3) The atoms in O_2 are joined by a *double* bond.

18. **1** Refer to the explanation given in Question 5 and use the temperature equation in Reference Table *T*:

$$K = °C + 273$$
$$°C = K - 273$$
$$°C = 273\ K - 273 = 0°C$$

19. **4** Refer to Reference Table *R*. Alcohols are organic compounds containing the –OH functional group and have names that end in –ol. Methanol is an alcohol containing one carbon atom and has the condensed structural formula CH_3OH.

20. **3** In the process of saponification, a fat (that is, an ester consisting of glycerol and fatty acids) is hydrolyzed by a base such as NaOH or KOH. The products of this reaction are glycerol and the sodium (or potassium) salts of the fatty acid. Collectively, these salts are known as soap.

Wrong Choices Explained:

(1) In the process of addition, a substance such as Br_2 is added to an alkene or an alkyne. As a result, the multiple bond is broken.

(2) In the process of substitution, a substance such as Cl_2 is reacted with an alkane. As a result, one or more hydrogen atoms are replaced by Cl atoms.

(4) In the process of polymerization, smaller molecules (the monomers) are joined to form a larger molecule (the polymer).

21. **4** Transmutation is defined as the nuclear process in which one atom is changed into another. Transmutation may occur spontaneously, or it may be induced.

22. **3** Reduction is defined as the process in which electrons are gained. As a result, the oxidation number of the substance that is reduced becomes more negative: that is, it decreases.

23. **2** Refer to the Periodic Table of the Elements. An Na^+ ion is formed by removing one electron from Na; its electron configuration is $(2–8–1) – 1 = 2–8$, which is the electron configuration of Ne, a noble gas.

24. **1** The term "fusion" is a synonym for melting. The heat of fusion is the amount of heat energy needed to change a given amount of solid to liquid at the melting point of the solid. Of the choices given, only choice (1) is an equation that represents a transition between solid and liquid.

Wrong Choices Explained:

(2) The heat term in this reaction is commonly known as the *heat of neutralization*.

(3) The heat term in this reaction is commonly known as the *heat of vaporization*.

25. **2** An Arrhenius acid is defined as a substance that produces H^+ ion as the only positive ion in aqueous solution. Refer to Reference Table *K* and note that HBr is very similar in structure to the Arrhenius acid HCl.

26. **4** The penetrating power of an emitted particle is measured by the thickness of a given material that is needed to stop that particle. The order of penetrating power, from highest to lowest, is:

$$gamma > beta \approx positron > alpha$$

27. **1** Refer to Reference Table *J*. The elements in Group 17 are the most active nonmetals. Chemical reactivity is related to the ease with which an atom gains electrons. As the atomic number increases, the reactivity of the element decreases.

28. **1** A liquid boils when its vapor pressure equals the atmospheric pressure above the liquid. At a lower atmospheric pressure, the vapor pressure of the liquid will reach the atmospheric pressure at a lower temperature. Refer to Reference Table *H*: At an atmospheric pressure of 101.3 kPa, the boiling point of ethanol is approximately 78°C; at an atmospheric pressure of 70 kPa, the boiling point of ethanol is approximately 69°C.

29. **3** Oxidation involves the loss of *electrons*: the number of neutrons in the nucleus remains unaffected by the process.

30. **2** Refer to Reference Table *H*. The vapor pressure of all liquids increases with increasing temperature.

PART B–1

31. **3** As the atomic number increases across Period 2, the nuclear charge increases and successive electrons are added to the same valence shell. The increased electrical attraction reduces the radius of each succeeding atom. Since metals lie at the left of the period and nonmetals at the right, the radii of the metals are generally larger than the radii of the nonmetals. This can be verified by referring to Reference Table S.

Wrong Choices Explained:
(1), (2), (4) Refer to Reference Table S. Metallic elements have *smaller* ionization energies, electronegativities, and atomic numbers than nonmetallic elements. These are direct results of the increases in nuclear charge and the decreases in atomic radius.

32. **4** Refer to Reference Table S. As the atomic number of each successive element within Group 2 increases, the atomic radius also increases. As a result, the energy needed to remove the most loosely held electron in each successive atom *decreases*.

33. **1** The boiling point of a liquid is directly related to the intermolecular forces present within the liquid. Since HF has the highest boiling point, it has the strongest intermolecular forces.

34. **2** Use Reference Table I to determine which salt has a negative ΔH when it dissolves. Of the choices given, only choice (2), LiBr, releases energy when it dissolves in water.

35. **2** The appearance of a yellow color could be determined only by direct observation.

36. **2** Use Equation 5B on Reference Table T. Remember that 200 mL = 0.2 L.

$$\text{molarity} = \frac{\text{moles of solute}}{\text{liters of solution}}$$

$$\text{moles of solute} = \text{molarity} \cdot \text{liters of solution}$$

$$\text{moles of solute} = (1 \text{ M}) \cdot (0.2 \text{ L}) = \textbf{0.2 mol}$$

Wrong Choice Explained:
(4) You forgot to convert milliliters to liters.

37. **4** Increasing the temperature increases the average kinetic energy of the reacting molecules. As a result, both the total number of collisions and the number of effective collisions between reacting molecules increase.

Wrong Choices Explained:
(1), (2) The activation energy of a reaction is changed only in the presence of a catalyst; it does not depend on the temperature at which the reaction occurs.

38. **3** According to LeChâtelier's principle, an increase in temperature favors the endothermic reaction of the equilibrium system. Since the forward reaction is endothermic, the equilibrium will shift to the right, leading to an increase in [HI].

39. **2** The entropy of a system is a measure of its disorder. Entropy depends on a number of factors, including phase (solid < liquid < gas). Choice (2), $KNO_3(s)$, is the only compound that is a solid.

40. **3** Use the solubility guidelines found in Reference Table F. As a class, the hydroxides form mostly insoluble compounds.

41. **1** Use Reference Table I. The formation of liquid H_2O from its elements is exothermic. Therefore, the potential energy of the products will be less than the potential energy of the reactants. Only the diagram given in choice (1) represents an exothermic reaction.

Wrong Choice Explained:
(2) The diagram represents an endothermic reaction because the potential energy of the products is greater than the potential energy of the reactants.

42. **4** Carbon can form a maximum of four bonds with neighboring atoms. In choice (4), the second carbon from the left contains *five* bonds.

43. **4** To balance a nuclear equation, you must be certain that the atomic numbers (the subscripts) and the mass numbers (the superscripts) are equal on both sides of the equation. Apply this rule to the given equation to obtain:

$$\text{$^2_1$H} + \text{$^2_1$H} \rightarrow \text{$^4_2$X} + \text{energy}$$

Refer to Reference Table O, which identifies X as an alpha particle, ^4_2He.

44. **4** The normal boiling point of a liquid is the temperature at which the liquid boils under a pressure of 1 atmosphere (101.3 kPa). Since the boiling point increases with increasing atmospheric pressure, it follows that the normal boiling point of the liquid must be *greater* than 60°C.

45. **2** Use Equation 7 on Reference Table *T*:

$$M_A V_A = M_B V_B$$
$$M_A \cdot 50.\ \text{mL} = 0.50\ \text{M} \cdot 150\ \text{mL}$$
$$M_A = \textbf{1.5 M}$$

46. **3** The number of protons in the atom of an element is given by its atomic number. When moving across Period 3, from left to right, the atomic numbers increase uniformly from 11 through 18.

Wrong Choices Explained:
(1), (2) Across Period 2, from left to right, both electronegativity and atomic mass increase.
(4) Across Period 2, from left to right, metallic character decreases.

47. **2** The balanced equation is:

$$2\,\text{Al} + 3\,\text{CuSO}_4 \rightarrow 1\,\text{Al}_2(\text{SO}_4)_3 + 3\,\text{Cu}$$

48. **2** According to Reference Table *G*, the solubility of NH_4Cl is 52 g per 100 g H_2O at 50°C and 35 g per 100 g H_2O at 10°C. When cooled from 50°C to 10°C, 17 grams (52 g − 35 g) of NH_4Cl will precipitate.

49. **3** Use Equations 5B and 2 on Reference Table *T*. Refer to the Periodic Table of the Elements to determine that the gram-formula mass of NaI is 150 grams per mole.

$$\text{molarity} = \frac{\text{moles of solute}}{\text{liters of solution}}$$
$$\text{moles of solute} = \text{molarity} \cdot \text{liters of solution}$$
$$\text{moles of solute} = (0.010\ \text{M}) \cdot (1.0\ \text{L}) = 0.010\ \text{mol}$$

$$\text{number of moles} = \frac{\text{mass}}{\text{gram-formula mass}}$$

$$0.010 \text{ mol} = \frac{\text{mass}}{150 \text{ g/mol}} = \mathbf{1.5 \text{ g}}$$

50. **1** *As the equation is written,* $\Delta H = -571.6$ kJ for this *exothermic* reaction. Since 571.6 kilojoules of energy are released when 2 moles of H_2O water are formed, the formation of 1 mole will release one-half that amount, 285.8 kilojoules. Therefore, $\Delta H = -285.8$ kJ.

PART B–2

[Point values are indicated in brackets.]

51. Refer to the Periodic Table of the Elements. Lead (Pb) has oxidation numbers of 2+ and 4+; sulfur (S) has a *negative* oxidation number of 2–. Therefore, lead and sulfur can form two compounds: PbS and PbS_2.

Student *X* claims that the rock samples contain two different mixtures of lead and sulfur. There are a number of possibilities. The rocks could contain two different amounts of *uncombined* lead and sulfur atoms. The rocks could contain different quantities of the two lead sulfides. The rocks could contain two different mixtures of combined and uncombined elements. Since lead is represented with a closed circle (●) and sulfur with an open circle (○), we can represent student *X*'s explanation with the following possible particle diagrams:

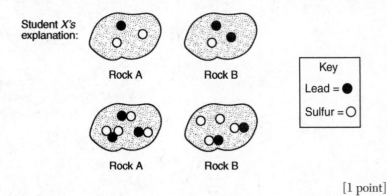

[1 point]

Student *Y* claims that the rock samples contain two different compounds of lead. A possible particle diagram that supports student *Y*'s explanation is:

Student *Y's* explanation:

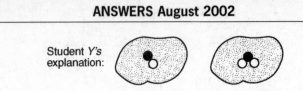

Rock A   Rock B   [1 point]

52. Refer to the Periodic Table of the Elements. The electron configuration of Na is 2–8–1; the electron configuration of K is 2–8–8–1. When these atoms form ions by losing a single electron each, the electron configuration of Na^+ is 2–8, and the electron configuration of K^+ is 2–8–8. Since Na^+ has only two shells while K^+ has three shells, the nucleus of the Na^+ ion is more effective in drawing the outer shell of electrons closer to the nucleus than is the K^+ ion.   [2 points]

53. **a** Possible diagrams for calcium oxide include:

$$Ca^{2+} [\, :\!\ddot{\underset{..}{O}}\!: \,]^{\,2-}$$

$$Ca \overset{\rightarrow}{\rightarrow} :\!\ddot{\underset{..}{O}}\!:$$

$$Ca \quad {}^{x}_{x}\!\ddot{\underset{..}{O}}\!:$$ [1 point]

b Possible diagrams for hydrogen bromide include:

$$H : \ddot{\underset{..}{Br}} :$$

$$H - \ddot{\underset{..}{Br}} :$$ [1 point]

c Possible diagrams for carbon dioxide include:

$$:\!\dot{\underset{.}{O}} = C = \dot{\underset{.}{O}}\!:$$

$$:\!\ddot{O}::C::\ddot{O}\!:$$ [1 point]

54. **a** When a phase change occurs, the temperature remains constant; temperature changes occur only in regions in which a single phase exists. This is shown in the diagram below:

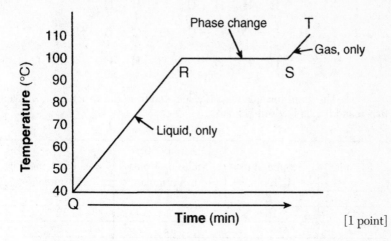

[1 point]

b As the temperature rises over region QR, the average kinetic energy of the molecules increases; that is, the molecules, on the average, move faster. [1 point]

c As heat is added over region RS, the potential energy of the molecules increases as it changes from liquid to gas. [1 point]

55. Isomers of a compound have the same molecular formula but different spatial arrangements. The diagram shows the structural formula for *n*-butane. The second isomer of butane, known as isobutane or 2-methylpropane, is shown below:

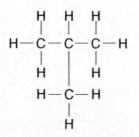

[1 point]

56. **a** The ester ethyl butanoate is derived from ethanol (C_2H_5OH) and butanoic acid (C_3H_7COOH). The structural formula of the ester is shown below:

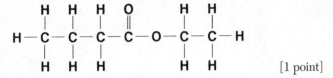

[1 point]

b The gram-formula mass is calculated from the structural formula in part **a** and the table given below:

Element	Atomic Mass (g/mol)	Number of Moles of Atoms in Formula	Mass of Element in Formula/g
C	12	6	72
O	16	2	32
H	1	12	12
		Formula mass =	**116**

[1 point]

57. **a** In this redox reaction, each Al atom is oxidized to Al^{3+} by losing 3 electrons. Acceptable answers include:

$$Al \rightarrow Al^{3+} + 3e^- \quad or$$
$$4Al \rightarrow 4Al^{3+} + 12e^- \quad or$$
$$Al - 3e^- \rightarrow Al^{3+} \quad or$$
$$4Al - 12e^- \rightarrow 4Al^{3+}$$

[1 point]

b The total charge of Al^{3+} in Al_2O_3 is 6+ [(2)·(3+)]. Therefore, the total charge of the oxygen must be 6− (Al_2O_3 is neutral). Since the 6− charge is carried by *three* oxide ions, the oxidation number of each ion is **2−**.   [1 point]

PART C

[Point values are indicated in brackets.]

58. **a** Advantages of using nuclear power rather than burning fossil fuels include less air pollution, lower cost, conserving the fossil-fuel reserves, and the greater amount of energy produced by nuclear power.   [1 point]

b The risks associated with the use of nuclear power include nuclear meltdown, contamination of the environment, biological hazards such as cancer and genetic mutations, the lack of adequate storage facilities for spent fuel rods, and the dangers of the radiation itself.   [1 point]

c The dangers associated with the incorporation of Sr-90, a beta emitter, into bone tissue include damage to surrounding soft tissues, bone cancer, and irradiation of red blood cells, leading to cancers of the blood.   [1 point]

59. **a** An electrolyte is a solution that conducts electricity when dissolved. Three of the flasks contain electrolytes:

- the ionic compound KCl produces K^+ and Cl^- in solution;
- the Arrhenius base $Ba(OH)_2$ produces Ba^{2+} and OH^- ions in solution;
- the Arrhenius acid CH_3COOH produces H^+ [H_3O^+] in solution.   [1 point]

b A low pH (<7) is associated with an acidic substance, such as CH_3COOH.   [1 point]

c Substances have a low pH when they are acidic: they yield an excess of H^+ [H_3O^+] in solution.   [1 point]

d An Arrhenius base, such as $Ba(OH)_2$ will react with an Arrhenius acid to form a salt and water.   [1 point]

e $Ba(OH)_2(aq)$ has the lowest freezing point because $Ba(OH)_2$ produces the greatest number of particles (ions) when dissolved in solution. [2 points]

60. **a** A particle diagram for the products $K^+(aq)$ and $NO_3^-(aq)$ is shown below:

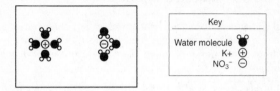

Note that the K^+ and the NO_3^- ions are separated.   [1 point]

Note that water molecules surround each ion. The oxygen atoms, which form the more negative end of the H_2O molecules, face toward the K^+ ion; the hydrogen atoms, which form the more positive ends of the H_2O molecules, face toward the NO_3^- ion.   [1 point]

b Since a saturated solution is an *equilibrium* system, the opposing rates (dissolving and crystallization) must be equal.   [1 point]

61. The graph of electron affinity versus atomic number is shown below:

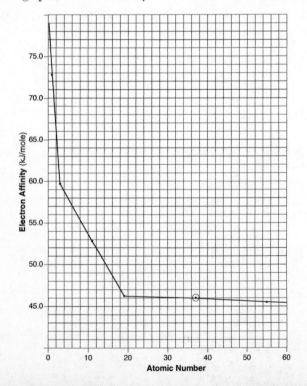

a Note that both axes are labeled and the scale is appropriate. (That is, the graph fills as much of the grid as possible.)   [1 point]

b Note that each data point is plotted correctly and that straight lines connect each pair of adjacent data points.   [1 point]

c The value of electron affinity corresponding to an atomic number of 37, as determined from the graph, has an acceptable range of 45.6–46.3 kilojoules per mole.   [1 point]

62. **a**

volume of sample = (volume of water and sample) − (voume of water)

volume of sample = (21.5 mL − 20.0 mL) = 1.5 mL

$$\text{density of sample} = \frac{\text{mass of sample}}{\text{volume of sample}}$$

$$= \frac{10.23 \text{ g}}{1.5 \text{ mL}} = \textbf{6.8 g/mL}$$

A total of 3 points is awarded for part **a**, distributed as follows:

- Setting up the equation properly.   [1 point]
- Answer is within the range of 6.7–6.9 g/mL and is expressed to *two* significant figures.   [1 point]
- Including units with the numerical answer.   [1 point]

b The percent error is calculated as follows:

$$\% \text{ error} = \frac{(\text{accepted value}) - (\text{measured value})}{(\text{accepted value})} \times 100\%$$

$$\% \text{ error} = \frac{(6.93 \text{ g/mL}) - (6.8 \text{ g/mL})}{(6.93 \text{ g/mL})} \times 100\%$$

$$= \frac{0.13 \text{ g/mL}}{6.93 \text{ g/mL}} = \textbf{1.9\%}$$

The acceptable range is 1.8%–2.0%.   [1 point]

c If the volume of the sample was determined *before* its mass was determined, the experimenter would be weighing a *wet* sample. The water clinging to the sample would introduce an error when measuring the mass. [1 point]

Mark (✓) the questions you answered correctly. Count the number of checks and follow the formulas given to determine your score on each topic.

Core Area	☐ Questions Answered Correctly

61a, 61b, 61c, 62a, 62b, 62c

Section M—Math Skills
☐ Number of checks ÷ 6 × 100 = _____%

___

1, 2

Section I—Atomic Concepts
☐ Number of checks ÷ 2 × 100 = _____%

___

3, 4, 15, 27, 31, 32, 46, 52

Section II—Periodic Table
☐ Number of checks ÷ 7 × 100 = _____%

___

6, 9, 47, 56b

Section III—Moles/Stoichiometry
☐ Number of checks ÷ 4 × 100 = _____%

___

10, 11, 12, 17, 23, 53a, 53b, 53c, 60a

Section IV—Chemical Bonding
☐ Number of checks ÷ 9 × 100 = _____%

___

5, 7, 14, 16, 24, 28, 30, 33, 34, 35, 36, 40, 44, 48, 49, 50, 51, 54a, 54b, 54c, 59e

Section V—Physical Behavior of Matter
☐ Number of checks ÷ 21 × 100 = _____%

___

13, 18, 37, 38, 39, 41, 60b

Section VI—Kinetics and Equilibrium
☐ Number of checks ÷ 7 × 100 = _____%

___

19, 20, 42, 55, 56a

Section VII—Organic Chemistry
☐ Number of checks ÷ 5 × 100 = _____%

___

22, 29, 57a, 57b

Section VIII—Oxidation–Reduction
☐ Number of checks ÷ 4 × 100 = _____%

___

25, 45, 59a, 59b, 59c, 59d

Section IX—Acids, Bases, and Salts
☐ Number of checks ÷ 6 × 100 = _____%

___

8, 21, 26, 43, 58a, 58b, 58c

Section X—Nuclear Chemistry
☐ Number of checks ÷ 7 × 100 = _____%

Examination January 2003
Physical Setting/Chemistry

PART A

Answer all questions in this part.

Directions (1-30): For *each* statement or question, write in the answer space the *number* of the word or expression that, of those given, best completes the statement or answers the question. Some questions may require the use of the *Reference Tables for Physical Setting/Chemistry*.

1 Which statement best describes electrons?

(1) They are positive subatomic particles and are found in the nucleus.

(2) They are positive subatomic particles and are found surrounding the nucleus.

(3) They are negative subatomic particles and are found in the nucleus.

(4) They are negative subatomic particles and are found surrounding the nucleus.

1 _____

2 During a flame test, ions of a specific metal are heated in the flame of a gas burner. A characteristic color of light is emitted by these ions in the flame when the electrons

   (1) gain energy as they return to lower energy levels
   (2) gain energy as they move to higher energy levels
   (3) emit energy as they return to lower energy levels
   (4) emit energy as they move to higher energy levels   2_____

3 In which list are the elements arranged in order of increasing atomic mass?

   (1) Cl, K, Ar   (3) Te, I, Xe
   (2) Fe, Co, Ni   (4) Ne, F, Na   3_____

4 In which compound does chlorine have the highest oxidation number?

   (1) NaClO   (3) NaClO$_3$
   (2) NaClO$_2$ (4) NaClO$_4$ 4_____

5 Which event must *always* occur for a chemical reaction to take place?

   (1) formation of a precipitate
   (2) formation of a gas
   (3) effective collisions between reacting particles
   (4) addition of a catalyst to the reaction system   5_____

6 Which Group of the Periodic Table contains atoms with a stable outer electron configuration?

   (1) 1   (3) 16
   (2) 8   (4) 18   6_____

7 From which of these atoms in the ground state can a valence electron be removed using the *least* amount of energy?

(1) nitrogen   (3) oxygen

(2) carbon   (4) chlorine   7____

8 What is the percent by mass of oxygen in H_2SO_4? [formula mass = 98]

(1) 16%   (3) 65%

(2) 33%   (4) 98%   8____

9 An atom of carbon-12 and an atom of carbon-14 differ in

(1) atomic number

(2) mass number

(3) nuclear charge

(4) number of electrons   9____

10 The strength of an atom's attraction for the electrons in a chemical bond is the atom's

(1) electronegativity   (3) heat of reaction

(2) ionization energy   (4) heat of formation   10____

11 Which type or types of change, if any, can reach equilibrium?

(1) a chemical change, only

(2) a physical change, only

(3) both a chemical and a physical change

(4) neither a chemical nor a physical change   11____

12 An increase in the average kinetic energy of a sample of copper atoms occurs with an increase in

(1) concentration   (3) pressure

(2) temperature   (4) volume   12____

13 The empirical formula of a compound is CH_2. Which molecular formula is correctly paired with a structural formula for this compound?

(1) C_2H_4 H—C—C—H
                     | |
                     H   H

(2) C_2H_4 H—C=C—H
                     | |
                     H   H

(3) C_3H_8 H—C—C—C—H
                     H   H   H (above)
                     H   H   H (below)

(4) C_3H_8 H—C=C—C—H
                     H   H   H (above)
                     H   H   H (below)

13_____

14 Given the equation:

$$:\overset{..}{\underset{..}{F}}\cdot + 1e^- \longrightarrow \left[:\overset{..}{\underset{..}{F}}:\right]^-$$

This equation represents the formation of a

(1) fluoride ion, which is smaller in radius than a fluorine atom
(2) fluoride ion, which is larger in radius than a fluorine atom
(3) fluorine atom, which is smaller in radius than a fluoride ion
(4) fluorine atom, which is larger in radius than a fluoride ion

14_____

15 The high electrical conductivity of metals is primarily due to

(1) high ionization energies
(2) filled energy levels
(3) mobile electrons
(4) high electronegativities

15____

16 One similarity between all mixtures and compounds is that both

(1) are heterogeneous
(2) are homogeneous
(3) combine in a definite ratio
(4) consist of two or more substances

16____

17 Which phase change results in the release of energy?

(1) $H_2O(s) \rightarrow H_2O(\ell)$
(2) $H_2O(s) \rightarrow H_2O(g)$
(3) $H_2O(\ell) \rightarrow H_2O(g)$
(4) $H_2O(g) \rightarrow H_2O(\ell)$

17____

18 Which compound has an isomer?

18____

19 What occurs when NaCl(s) is added to water?

(1) The boiling point of the solution increases, and the freezing point of the solution decreases.

(2) The boiling point of the solution increases, and the freezing point of the solution increases.

(3) The boiling point of the solution decreases, and the freezing point of the solution decreases.

(4) The boiling point of the solution decreases, and the freezing point of the solution increases.   19____

20 Which radioisotope is a beta emitter?

(1) ^{90}Sr   (3) ^{37}K

(2) ^{220}Fr   (4) ^{238}U   20____

21 When a mixture of water, sand, and salt is filtered, what passes through the filter paper?

(1) water, only

(2) water and sand, only

(3) water and salt, only

(4) water, sand, and salt   21____

22 A hydrate is a compound that includes water molecules within its crystal structure. During an experiment to determine the percent by mass of water in a hydrated crystal, a student found the mass of the hydrated crystal to be 4.10 grams. After heating to constant mass, the mass was 3.70 grams. What is the percent by mass of water in this crystal?

(1) 90.%   (3) 9.8%

(2) 11%   (4) 0.40%   22____

23 Which of these 1 M solutions will have the highest pH?

(1) NaOH   (3) HCl

(2) CH_3OH (4) NaCl   23____

24 Which physical property makes it possible to separate the components of crude oil by means of distillation?

(1) melting point   (3) solubility
(2) conductivity   (4) boiling point   24_____

25 In saturated hydrocarbons, carbon atoms are bonded to each other by

(1) single covalent bonds, only
(2) double covalent bonds, only
(3) alternating single and double covalent bonds
(4) alternating double and triple covalent bonds   25_____

26 Which formula correctly represents the product of an addition reaction between ethene and chlorine?

(1) CH_2Cl_2 (3) $C_2H_4Cl_2$
(2) CH_3Cl (4) C_2H_3Cl 26_____

27 When a neutral atom undergoes oxidation, the atom's oxidation state

(1) decreases as it gains electrons
(2) decreases as it loses electrons
(3) increases as it gains electrons
(4) increases as it loses electrons   27_____

28 Given the equation:

$$C(s) + H_2O(g) \rightarrow CO(g) + H_2(g)$$

Which species undergoes reduction?

(1) $C(s)$ (3) C^{2+}
(2) H^+ (4) $H_2(g)$ 28_____

29 Which equation is an example of artificial transmutation?

(1) $^{9}_{4}Be + ^{4}_{2}He \rightarrow ^{12}_{6}C + ^{1}_{0}n$
(2) $U + 3 F_2 \rightarrow UF_6$
(3) $Mg(OH)_2 + 2 HCl \rightarrow 2 H_2O + MgCl_2$
(4) $Ca + 2 H_2O \rightarrow Ca(OH)_2 + H_2$

29____

30 Which species can conduct an electric current?

(1) $NaOH(s)$ (3) $H_2O(s)$
(2) $CH_3OH(aq)$ (4) $HCl(aq)$

30____

PART B–1

Answer all questions in this part.

Directions (31–50): For *each* statement or question, write in the answer space the *number* of the word or expression that, of those given, best completes the statement or answers the question. Some questions may require the use of the *Reference Tables for Physical Setting/Chemistry.*

31 According to Table *N*, which radioactive isotope is best for determining the actual age of Earth?

(1) ^{238}U (3) ^{60}Co
(2) ^{90}Sr (4) ^{14}C

31____

32 Given the following solutions:

Solution *A*: pH of 10
Solution *B*: pH of 7
Solution *C*: pH of 5

Which list has the solutions placed in order of increasing H^+ concentration?

(1) *A, B, C*   (3) *C, A, B*
(2) *B, A, C*   (4) *C, B, A*

32____

33 Which statement explains why nuclear waste materials may pose a problem?

(1) They frequently have short half-lives and remain radioactive for brief periods of time.
(2) They frequently have short half-lives and remain radioactive for extended periods of time.
(3) They frequently have long half-lives and remain radioactive for brief periods of time.
(4) They frequently have long half-lives and remain radioactive for extended periods of time.

33_____

34 A compound whose water solution conducts electricity and turns phenolphthalein pink is

(1) HCl
(2) $HC_2H_3O_2$
(3) NaOH
(4) CH_3OH

34_____

35 Which of the following solids has the highest melting point?

(1) $H_2O(s)$
(2) $Na_2O(s)$
(3) $SO_2(s)$
(4) $CO_2(s)$

35_____

36 Hydrogen has three isotopes with mass numbers of 1, 2, and 3 and has an average atomic mass of 1.00794 amu. This information indicates that

(1) equal numbers of each isotope are present
(2) more isotopes have an atomic mass of 2 or 3 than of 1
(3) more isotopes have an atomic mass of 1 than of 2 or 3
(4) isotopes have only an atomic mass of 1

36_____

37 Which list of elements contains *two* metalloids?

(1) Si, Ge, Po, Pb
(2) As, Bi, Br, Kr
(3) Si, P, S, Cl
(4) Po, Sb, I, Xe

37_____

38 Given the reaction:

$$S(s) + O_2(g) \rightarrow SO_2(g) + energy$$

Which diagram best represents the potential energy changes for this reaction?

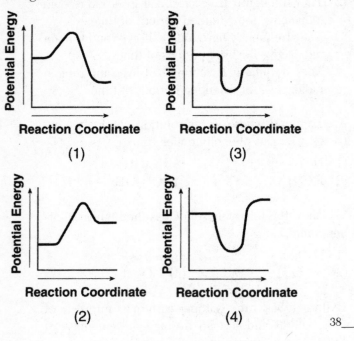

38____

39 A chemist performs the same tests on two homogeneous white crystalline solids, A and B. The results are shown in the table below.

	Solid A	Solid B
Melting Point	High, 801°C	Low, decomposes at 186°C
Solubility in H_2O (grams per 100.0 g H_2O at 0°C)	35.7	3.2
Electrical Conductivity (in aqueous solution)	Good conductor	Nonconductor

The results of these tests suggest that

(1) both solids contain only ionic bonds
(2) both solids contain only covalent bonds
(3) solid A contains only covalent bonds and solid B contains only ionic bonds
(4) solid A contains only ionic bonds and solid B contains only covalent bonds

39_____

40 Solubility data for four different salts in water at 60°C are shown in the table below.

Salt	Solubility in Water at 60°C
A	10 grams / 50 grams H_2O
B	20 grams / 60 grams H_2O
C	30 grams / 120 grams H_2O
D	40 grams / 80 grams H_2O

Which salt is most soluble at 60°C?

(1) A   (3) C
(2) B   (4) D

40_____

41 Which phase change represents a *decrease* in entropy?

(1) solid to liquid   (3) liquid to gas

(2) gas to liquid   (4) solid to gas   41____

42 Given the equation:

$$2\ C_2H_2(g) + 5\ O_2(g) \rightarrow 4\ CO_2(g) + 2\ H_2O(g)$$

How many moles of oxygen are required to react completely with 1.0 mole of C_2H_2?

(1) 2.5   (3) 5.0

(2) 2.0   (4) 10   42____

43 A student intended to make a salt solution with a concentration of 10.0 grams of solute per liter of solution. When the student's solution was analyzed, it was found to contain 8.90 grams of solute per liter of solution. What was the percent error in the concentration of the solution?

(1) 1.10%   (3) 11.0%

(2) 8.90%   (4) 18.9%   43____

44 What is the molarity of a solution of NaOH if 2 liters of the solution contains 4 moles of NaOH?

(1) 0.5 M   (3) 8 M

(2) 2 M   (4) 80 M   44____

45 A gas occupies a volume of 40.0 milliliters at 20°C. If the volume is increased to 80.0 milliliters at constant pressure, the resulting temperature will be equal to

(1) $20°C \times \dfrac{80.0\ mL}{40.0\ mL}$ (3) $293\ K \times \dfrac{80.0\ mL}{40.0\ mL}$

(2) $20°C \times \dfrac{40.0\ mL}{80.0\ mL}$ (4) $293\ K \times \dfrac{40.0\ mL}{80.0\ mL}$ 45____

46 According to Reference Table *J*, which of these metals will react most readily with 1.0 M HCl to produce $H_2(g)$?

(1) Ca   (3) Mg

(2) K   (4) Zn   46____

47 The graph below represents the heating curve of a substance that starts as a solid below its freezing point.

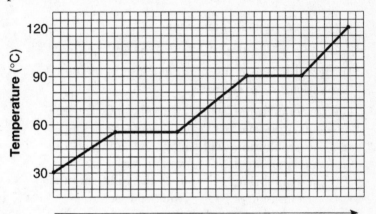

Time (minutes)

What is the melting point of this substance?

(1) 30°C   (3) 90°C

(2) 55°C   (4) 120°C   47____

48 Given the unbalanced equation:

$$\_\_\_Fe_2O_3 + \_\_\_CO \rightarrow \_\_\_Fe + \_\_\_CO_2$$

When the equation is correctly balanced using the *smallest* whole-number coefficients, what is the coefficient of CO?

(1) 1   (3) 3

(2) 2   (4) 4   48____

49 Which type of organic compound is represented by the structural formula shown below?

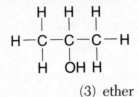

(1) aldehyde   (3) ether
(2) alcohol   (4) ester   49_____

50 Given the system at equilibrium:

$$N_2O_4(g) + 58.1 \text{ kJ} \rightleftharpoons 2\ NO_2(g)$$

What will be the result of an increase in temperature at constant pressure?

(1) The equilibrium will shift to the left, and the concentration of $NO_2(g)$ will decrease.
(2) The equilibrium will shift to the left, and the concentration of $NO_2(g)$ will increase.
(3) The equilibrium will shift to the right, and the concentration of $NO_2(g)$ will decrease.
(4) The equilibrium will shift to the right, and the concentration of $NO_2(g)$ will increase.   50_____

PART B–2

Answer all questions in this part.

Directions (51-61): Record your answers on the answer sheet provided in the back. Some questions may require the use of the *Reference Tables for Physical Setting/Chemistry.*

51 In the boxes provided *on your answer sheet:*

  a Draw *two* different compounds, one in each box, using the representations for atoms of element *X* and element *Z* given below.   [1]

  Atom of element *X* = O
  Atom of element Z = ●

  b Draw a mixture of these two compounds.   [1]

52 At equilibrium, nitrogen, hydrogen, and ammonia gases form a mixture in a sealed container. The data table below gives some characteristics of these substances.

Data Table

Gas	Boiling Point	Melting Point	Solubility in Water
Nitrogen	−196°C	−210°C	insoluble
Hydrogen	−252°C	−259°C	insoluble
Ammonia	−33°C	−78°C	soluble

  Describe how to separate ammonia from hydrogen and nitrogen.   [1]

Base your answers to questions 53 through 55 on the diagram of a voltaic cell provided *on your answer sheet* and on your knowledge of chemistry.

53 On the diagram provided *on your answer sheet,* indicate with one or more arrows the direction of electron flow through the wire.   [1]

54 Write an equation for the half-reaction that occurs at the zinc electrode.   [1]

55 Explain the function of the salt bridge.   [1]

---

56 Given the nuclear equation:

$$^{235}_{92}U + {}^{1}_{0}n \rightarrow {}^{142}_{56}Ba + {}^{91}_{36}Kr + 3\,{}^{1}_{0}n + energy$$

   a State the type of nuclear reaction represented by the equation.   [1]
   b The sum of the masses of the products is slightly less than the sum of the masses of the reactants. Explain this loss of mass.   [1]
   c This process releases greater energy than an ordinary chemical reaction does. Name another type of nuclear reaction that releases greater energy than an ordinary chemical reaction.   [1]

Base your answers to questions 57 through 60 on the information below.

Each molecule listed below is formed by sharing electrons between atoms when the atoms within the molecule are bonded together.

Molecule A: Cl_2

Molecule B: CCl_4

Molecule C: NH_3

57 In the box provided *on your answer sheet,* draw the electron-dot (Lewis) structure for the NH_3 molecule.   [1]

58 Explain why CCl_4 is classified as a nonpolar molecule.   [1]

59 Explain why NH_3 has stronger intermolecular forces of attraction than Cl_2.   [1]

60 Explain how the bonding in KCl is different from the bonding in molecules A, B, and C.   [1]

61 How is the bonding between carbon atoms different in unsaturated hydrocarbons and saturated hydrocarbons?   [1]

PART C

Answer all questions in this part.

Directions (62-74): Record your answers on the answer sheet provided in the back. Some questions may require the use of the *Reference Tables for Physical Setting/Chemistry.*

Base your answers to questions 62 through 64 on the information and diagram below.

One model of the atom states that atoms are tiny particles composed of a uniform mixture of positive and negative charges. Scientists conducted an experiment where alpha particles were aimed at a thin layer of gold atoms.

Most of the alpha particles passed directly through the gold atoms. A few alpha particles were deflected from their straight-line paths. An illustration of the experiment is shown below.

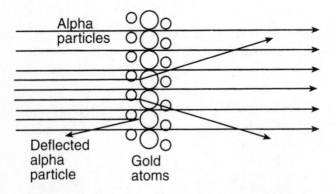

62 Most of the alpha particles passed directly through the gold atoms undisturbed. What does this evidence suggest about the structure of the gold atoms? [1]

63 A few of the alpha particles were deflected. What does this evidence suggest about the structure of the gold atoms? [1]

64 How should the original model be revised based on the results of this experiment? [1]

Base your answers to questions 65 through 67 on the information below.

When cola, a type of soda pop, is manufactured, $CO_2(g)$ is dissolved in it.

65 A capped bottle of cola contains $CO_2(g)$ under high pressure. When the cap is removed, how does pressure affect the solubility of the dissolved $CO_2(g)$? [1]

66 A glass of cold cola is left to stand 5 minutes at room temperature. How does temperature affect the solubility of the $CO_2(g)$? [1]

67 *a* In the space provided *on your answer sheet,* draw a set of axes and label one of them "Solubility" and the other "Temperature." [1]

 b Draw a line to indicate the solubility of $CO_2(g)$ versus temperature on the axes drawn in part *a*. [1]

Base your answers to questions 68 through 70 on the graph below, which shows the vapor pressure curves for liquids *A* and *B*.

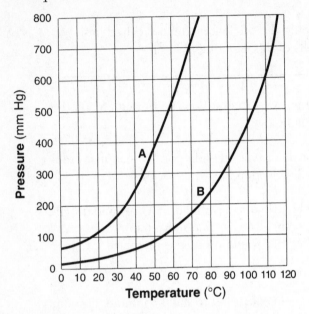

68. What is the vapor pressure of liquid *A* at 70°C? Your answer must include correct units. [2]

69. At what temperature does liquid *B* have the same vapor pressure as liquid *A* at 70°C? Your answer must include correct units. [2]

70. Which liquid will evaporate more rapidly? Explain your answer in terms of intermolecular forces. [2]

Base your answers to questions 71 through 74 on the information and data table below.

A titration setup was used to determine the unknown molar concentration of a solution of NaOH. A 1.2 M HCl solution was used as the titration standard. The following data were collected.

	Trial 1	Trial 2	Trial 3	Trial 4
Amount of HCl Standard Used	10.0 mL	10.0 mL	10.0 mL	10.0 mL
Initial NaOH Buret Reading	0.0 mL	12.2 mL	23.2 mL	35.2 mL
Final NaOH Buret Reading	12.2 mL	23.2 mL	35.2 mL	47.7 mL

71 Calculate the volume of NaOH solution used to neutralize 10.0 mL of the standard HCl solution in trial 3. Show your work. [2]

72 According to Reference Table M, what indicator would be most appropriate in determining the end point of this titration? Give one reason for choosing this indicator. [2]

73 Calculate the average molarity of the unknown NaOH solution for all four trials. Your answer must include the correct number of significant figures and correct units. [3]

74 Explain why it is better to use the average data from multiple trials rather than the data from a single trial to calculate the results of the titration. [1]

Answer Sheet
January 2003

Physical Setting/Chemistry

PART B–2

Answer Space

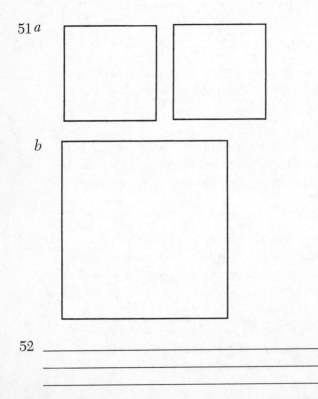

51 *a*

b

52 _____
_____
_____

53

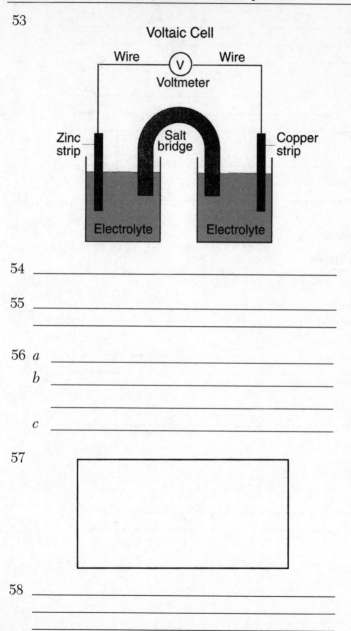

Voltaic Cell

Wire   (V)   Wire

Voltmeter

Zinc strip   Salt bridge   Copper strip

Electrolyte   Electrolyte

54 _____

55 _____
_____

56 *a* _____
  b _____
_____
  c _____

57

58 _____
_____
_____

59 _____

60 _____

61 _____

PART C

Answer Space

62 _____

63 _____

64 _____

65 _____

66 _____

67 *a* and *b*

68 _____

69 _____

70 Liquid: _____

   Explanation: _____

   _____

   _____

71

   _____**mL**

72 Indicator: _____

   Reason: _____

   _____

   _____

73

      _____

74 _____
    _____
    _____

Answers
January 2003
Physical Setting/Chemistry

Answer Key

PART A

1. 4	7. 2	13. 2	19. 1	25. 1
2. 3	8. 3	14. 2	20. 1	26. 3
3. 1	9. 2	15. 3	21. 3	27. 4
4. 4	10. 1	16. 4	22. 3	28. 2
5. 3	11. 3	17. 4	23. 1	29. 1
6. 4	12. 2	18. 4	24. 4	30. 4

PART B–1

31. 1	35. 2	39. 4	43. 3	47. 2
32. 1	36. 3	40. 4	44. 2	48. 3
33. 4	37. 1	41. 2	45. 3	49. 2
34. 3	38. 1	42. 1	46. 2	50. 4

Answers Explained

PART A

1. **4** The nucleus of an atom contains positively charged protons and neutrons, which have no charge. The negatively charged electrons are found surrounding the nucleus.

2. **3** When the ions of a specific metal are heated in a burner flame, electrons are raised to higher energy levels. As the electrons return to lower energy levels, photons with characteristic colors are emitted.

3. **1** Refer to the Periodic Table of the Elements. The respective atomic masses of Cl, K, and Ar are 35.453, 39.0983, and 39.948 atomic mass units.

4. **4** In all compounds, the sum of the oxidation numbers must equal 0. In all of the choices given in this question, sodium (Na) has an oxidation number of +1 and each oxygen (O) has an oxidation number of –2. In choice (4), $NaClO_4$, the Na accounts for +1 and the four O atoms account for –8. For the oxidation numbers to equal 0, chlorine (Cl) must have an oxidation number of +7.

Wrong Choices Explained:
(1) In NaClO, the oxidation number of Cl is +1.
(2) In $NaClO_2$, the oxidation number of Cl is +3.
(3) In $NaClO_3$, the oxidation number of Cl is +5.

5. **3** All reactions require collisions among the reacting particles. In order for a reaction to occur, the collisions must have sufficient energy and the particles must be oriented properly, that is, the collisions must be *effective*.

6. **4** The noble gases, found in Group 18, all have stable outer electron configurations. Helium contains two electrons in its first level, and the other elements all have eight electrons in their outer levels.

7. **2** Refer to Reference Table *S* and compare the ionization energies of each of the choices given:

Element	Ionization Energy (kJ/mol)
Nitrogen	1402
Carbon	1086
Oxygen	1314
Chlorine	1251

Since carbon has the smallest first ionization energy, it requires the least amount of energy to remove a valence electron.

8. **3** First, we complete the following table for H_2SO_4:

Element	Atomic Mass	Number of Atoms in Formula	Mass of Element in Formula
H	1	2	2
S	32	1	32
O	16	4	64
		Formula mass =	**98**

Since the contribution of oxygen to the formula mass is 64, we can calculate the percent composition by mass from the following relationship:

$$\frac{64}{98} \times 100\% = \mathbf{65\%}$$

9. **2** Isotopes of an atom are identified by their *mass numbers*: the number of protons and neutrons in their nuclei. An atom of carbon–12 contains a total of 12 protons and neutrons in its nucleus, and an atom of carbon–14 contains a total of 14 protons and neutrons in its nucleus.

Wrong Choices Explained:
(1) The atomic number of an atom is the number of protons in its nucleus.
(3) The nuclear charge of an atom is the total positive charge of its protons.
(4) All neutral atoms of a given element contain the same number of electrons.

10. **1** Electronegativity is *defined* as the strength of an atom's attraction for electrons in a chemical bond.

Wrong Choices Explained:

(2) The ionization energy of an element is the energy needed to remove an electron from the atom.

(3) The heat of reaction is the heat energy absorbed or liberated by a chemical reaction.

(4) The heat of formation is the heat energy absorbed or liberated when one mole of a compound is formed from its elements.

11. **3** A state of equilibrium exists when the rates of the forward and reverse processes are equal. For example, in the *chemical* change $N_2(g) + 3H_2(g) \rightleftharpoons 2NH_3(g)$, when the rate of NH_3 production equals the rate of NH_3 removal, a state of equilibrium exists. Similarly, in the *physical* change $H_2O(s) \rightleftharpoons H_2O(\ell)$, when the rate of the melting of ice equals the rate at which water freezes, a state of equilibrium exists. Therefore, both physical and chemical changes can reach equilibrium.

12. **2** The temperature of any substance is a measure of the average kinetic energy of its particles (atoms, molecules, or ions).

13. **2** The empirical formula of a compound is the result of reducing the molecular formula of a compound to *smallest whole numbers*. In this question, the molecular formula corresponding to the empirical formula CH_2 is C_2H_4. This eliminates choices (3) and (4). We must now inspect choices (1) and (2) to see which is correctly drawn. In choice (2), each carbon atom is correctly associated with four bonds and each hydrogen atom is correctly associated with a single bond.

Wrong Choice Explained:

(1) In this structure, each carbon atom is *incorrectly* associated with three bonds; its valence level is incomplete.

14. **2** On the left side of the equation we see the Lewis structure of a fluorine atom, and on the right side we see the structure of a fluoride ion. When an atom forms a negative ion, its radius *increases*.

15. **3** In a metallic substance, the conduction of an electric current is due to the presence of electrons in motion. Therefore, bonding in a metallic substance must give rise to mobile electrons.

Wrong Choices Explained:

(1), (2), (4) High ionization energies, filled energy levels, and high electronegativities are characteristic of *nonmetallic* substances.

16. **4** Mixtures and compounds must contain more than one substance. In compounds, however, the composition is fixed (as in water); in a mixture, the composition is variable (as in an ice cream soda).

17. **4** Refer to the table below, which shows the energy changes accompanying phase changes:

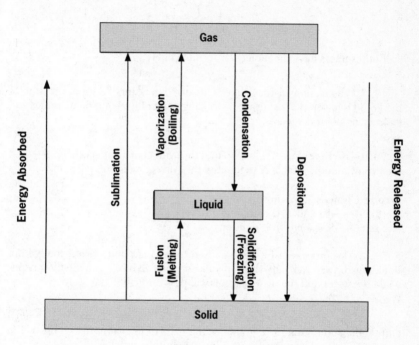

Of the choices given, only choice (4), $H_2O(g) \rightarrow H_2O(\ell)$, is accompanied by the release of energy.

18. **4** Two or more compounds are isomers if they have the same molecular formula but different structural formulas. Of the choices given, only choice (4),

has a structural isomer, which is shown below:

Both isomers have the molecular formula C_4H_{10}.

19. **1** When a nonvolatile solute is dissolved in water, the vapor pressure of the liquid is lowered. As a result, the boiling point of the solution increases and its freezing point decreases.

20. **1** Use Reference Table O to find the symbol of a beta particle (β^-), and then use Reference Table N to identify the correct radioisotope: ^{90}Sr.

Wrong Choices Explained:
(2), (4) ^{220}Fr and ^{238}U are *alpha* emitters.
(3) ^{37}K is a *positron* emitter.

21. **3** The process of filtration is used to separate undissolved solids from liquid mixtures. As a result, the undissolved sand will remain on the filter paper while the water and the dissolved salt will pass through it.

22. **3** The difference between the mass of the hydrated crystal (4.10 grams) and the heated crystal (3.70 grams) is the mass of the water driven off by the heating process. The percent water by mass in the crystal is:

$$\frac{4.10 \text{ g} - 3.70 \text{ g}}{4.10 \text{ g}} \times 100\% = \mathbf{9.8\%}$$

23. **1** A 1 M solution of NaOH is strongly basic and will have a pH of approximately 14, the highest pH of any of the other choices.

Wrong Choices Explained:
(3)   A 1 M solution of HCl is strongly acidic and will have a pH of approximately 0, the lowest pH of any of the other choices.
(2), (4)   1 M solutions of CH_3OH and NaCl are essentially neutral and will have a pH very close to 7.

24. **4** Crude oil is separated into its components by means of fractional distillation, a process that uses differences in boiling points to effect the separation.

25. **1** Saturated hydrocarbons are *defined* as compounds in which the carbon atoms are bonded to each other by means of single covalent bonds only.

26. **3** In the addition reaction between ethene and chlorine, the chlorine breaks the double bond in ethene and a chlorine atom bonds to each carbon atom. The reaction is shown below:

$$C_2H_4 + Cl_2 \rightarrow C_2H_4Cl_2$$

27. **4** Oxidation is defined as the *loss of electrons*. As a particle loses electrons, its oxidation number increases. The oxidation of a sodium atom to form a sodium ion is shown below:

$$Na^0 \rightarrow Na^+ + e^-$$

Note the oxidation number of Na increases from 0 to +1.

28. **2** When a particle undergoes reduction, its oxidation number *decreases*: Write the reaction with the oxidation numbers shown for each element:

$$C^0 + H_2^+O^{2-} \rightarrow C^{2+}O^{2-} + H_2^0$$

Observe that the H^+ in H_2O changes its oxidation number from +1 to 0 as it forms $H_2(g)$.

Wrong Choices Explained:
(1) C(s) is oxidized; its oxidation number increases.
(3), (4) C^{2+} and $H_2(g)$ are products of this reaction. Oxidation and reduction are defined only for the reactants.

29. **1** Artificial transmutation involves the use of a bombarding particle (such as $_2^4He$) to effect a nuclear reaction. In this reaction, an atom of Be–9 is bombarded by an alpha particle and is transmuted into an atom of C–12. In addition, a neutron is produced.

Wrong Choices Explained:
(2), (3), (4) These choices are not nuclear reactions, they are ordinary chemical changes that involve electrons.

30. **4** In order to conduct an electric current, *mobile charges* must be present. Metallic substances conduct electricity due to the presence of *mobile electrons*. If a liquid or a solution contains *mobile ions*, the liquid or the solution will also conduct electricity. In water, HCl ionizes nearly completely to form hydronium and chloride ions. Therefore, this solution conducts electricity:

$$HCl + H_2O \rightarrow H_3O^+ + Cl^-$$

Wrong Choices Explained:
(1) NaOH(s) contains ions, but they are not mobile. They are confined to the NaOH crystal.
(2), (3) These substances do not produce sufficient ions in solution to conduct electricity.

PART B–1

31. **1** Use Reference Table *N*. The age of an artifact is determined by using the half-life of a radioisotope contained within the artifact itself. In order to determine the age of very old artifacts, such as Earth itself, radioisotopes with very long half-lives must be used. Of the choices given, only choice (1), ^{238}U (half-life = 4.51×10^9 years), has a sufficiently long half-life.

32. **1** As the pH of a solution *decreases*, its H^+ concentration *increases*. Solution *A*, with a pH of 10, has the lowest H^+ concentration, and solution *C*, with a pH of 5, has the highest H^+ concentration.

33. **4** One of the major problems involving nuclear wastes is the long periods of time that they remain radioactive. This property is directly related to the long half-lives of these radioisotopes.

34. **3** Since the solution conducts electricity, it must contain ions. Since the solution turns phenolphthalein pink, it must be basic (see Reference Table *M*). Of the choices given, only choice (3), NaOH, meets both of these criteria.

Wrong Choices Explained:

(1) A solution of HCl conducts electricity, but it is acidic. Phenolphthalein would remain colorless.

(2) A solution of $HC_2H_3O_2$ also conducts electricity (very weakly), but it too is acidic. Phenolphthalein would remain colorless.

(4) A solution of CH_3OH would not conduct electricity.

35. **2** High melting points are found among ionic and covalent network solids. Choice (2), Na_2O, is an ionic solid.

Wrong Choices Explained:

(1), (3), (4) H_2O, SO_2, and CO_2 are molecular solids and have lower melting points.

36. **3** When the average atomic mass of an isotope is rounded, it indicates the mass number of the most abundant isotope. Since the average atomic mass is so close to 1, it indicates that hydrogen–1 must be the most abundant isotope.

37. **1** If we examine the abbreviated Periodic Table of the Elements shown below, we see that silicon (Si) and germanium (Ge) are classified as *metalloids* and have properties common to both metals and nonmetals.

GROUP

13	14	15	16	17
B	C	N	O	F
Al	Si	P	S	Cl
Ga	Ge	As	Se	Br
In	Sn	Sb	Te	I
Tl	Pb	Bi	Po	At

KEY

nonmetal
metalloid
metal

Wrong Choices Explained:

(2), (3), (4) Each of these lists contains only *one* metalloid (As, Si, and Sb, respectively).

38. **1** The reaction is exothermic (releases energy). As a consequence, the potential energy of the products is less than the potential energy of the reactants. The diagram below shows a labeled potential energy diagram for an exothermic reaction:

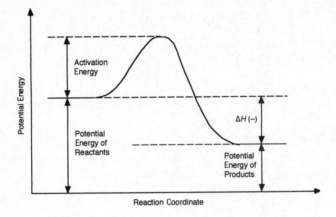

The diagram in choice (1) is most similar to this diagram.

39. **4** Solid *A* has a high melting point and good solubility in water. Additionally, its aqueous solution conducts electricity. These properties are characteristic of ionic compounds. Solid *B* has a low melting point and low solubility in water. Additionally, its aqueous solution does not conduct electricity. These properties are consistent with those of a molecular solid, that is, one that contains covalent bonds.

40. **4** The trick here is to convert the solubilities to a form that contains the same mass of H_2O for each salt. This is done in the table below in which the solubilities are expressed in grams of salt per 100 grams of H_2O:

Salt	Solubility in Water at 60°C	Solubility in Water at 60°C
A	10 g / 50 g H_2O →	20 g / 100 g H_2O
B	20 g / 60 g H_2O →	33 g / 100 g H_2O
C	30 g / 120 g H_2O →	25 g / 100 g H_2O
D	40 g / 80 g H_2O →	50 g / 100 g H_2O

The table clearly shows that salt *D* has the greatest solubility in water at 60°C.

41. **2** Entropy is the measure of disorder in a substance. A gas has a higher entropy than a liquid, and a liquid has a higher entropy than a solid. Of the phase changes given, only choice (2), gas to liquid, will be accompanied by a decrease in entropy.

42. **1** The coefficients of the equation represent the relative number of moles that react and are formed: 5 moles of oxygen (O_2) will react with 2 moles of C_2H_2. Use the factor-label method as shown below:

$$1.0 \text{ mol } C_2H_2 \cdot \frac{5 \text{ mol } O_2}{2 \text{ mol } C_2H_2} = \textbf{2.5 mol } O_2$$

43. **3** Use Equation 3 on Reference Table *T*:

$$\% \text{ error} = \frac{\text{measured value} - \text{accepted value}}{\text{accepted value}} \times 100$$

$$= \frac{8.90 \text{ g/L} - 10.0 \text{ g/L}}{10.0 \text{ g/L}} \times 100 = \textbf{11.0\%}$$

Note that the negative sign associated with the answer is ignored.

44. **2** Use Equation 5B on Reference Table *T*:

$$\text{molarity} = \frac{\text{moles of solute}}{\text{liters of solution}}$$

$$= \frac{4 \text{ mol}}{2 \text{ L}}$$

$$= \textbf{2 M}$$

45. **3** Use Equations 9 and 6 on Reference Table *T*:

$$K = {}^\circ C + 273$$

$$= 20{}^\circ C + 273$$

$$= \textbf{293 K}$$

At constant pressure, the combined gas law reduces to:

$$\frac{V_1}{T_1} = \frac{V_2}{T_2}$$

$$\frac{40.0 \text{ mL}}{293 \text{ K}} = \frac{80.0 \text{ K}}{T_2}$$

$$T_2 = \mathbf{293 \text{ K} \times \frac{80.0 \text{ mL}}{40.0 \text{ mL}}}$$

Wrong Choices Explained:
  (1) You forgot to convert the temperature from Celsius to Kelvin.
  (4) In solving for the temperature, you mistakenly inverted the volumes.

46. **2** Any metal found *above* H_2 on Reference Table *J* will react with H^+ (as in HCl) to produce H_2. The higher up on the chart the metal is, the more readily it will react with H^+. Of the choices given, choice (2), K, is highest on the chart.

47. **2** Refer to the labeled graph below:

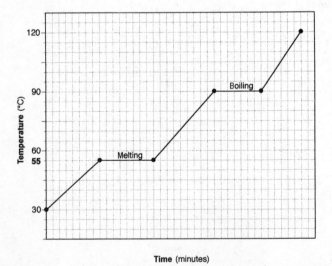

Each vertical box is 5°C. The temperature corresponding to melting is 55°C.

48. **3** The balanced equation (using smallest whole-number coefficients) is:

$$Fe_2O_3 + 3CO \rightarrow 2Fe + 3CO_2$$

49. **2** Refer to Reference Table R. The organic compound contains the hydroxyl (–OH) functional group and is classified as an alcohol.

50. **4** Increasing the temperature of a system at equilibrium will shift the system in the direction of the *endothermic* reaction (in this example, the *forward* reaction). Therefore, an increase in temperature will shift the equilibrium to the *right,* resulting in an increase of $NO_2(g)$.

PART B–2

[Point values are indicated in brackets.]

51. **a** You have the option of choosing which compounds to draw. However, the compounds *must* contain different numbers of X and Z, and different particles must be touching. Two possible examples are given below:

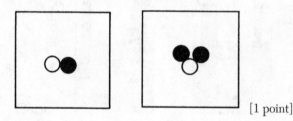

[1 point]

    b You must draw at least one of each particle you drew in part **a**. One possibility is shown below:

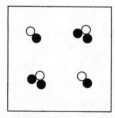

[1 point]

52. You need to describe only *one* technique for the separation of ammonia from the mixture. Three possibilities are given below:

- Since ammonia is soluble in water while hydrogen and nitrogen are not, the ammonia can be separated by passing the mixture of gases through water.
- Make use of the differences in the boiling points of the three gases. When the mixture of gases is cooled, ammonia will condense first since it has the highest boiling point.
- Convert all of the gases to liquids and *distill* the mixture.   [1 point]

53. Refer to the diagram below in which the arrows represent the direction of the electron flow through the wire:

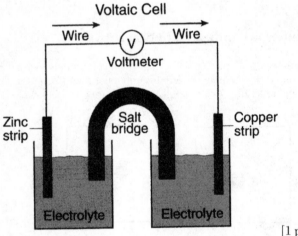

[1 point]

54. Zinc is *oxidized* at the zinc electrode. Its half-reaction is:

$$Zn^0 \rightarrow Zn^{2+} + 2e^- \ or$$
$$Zn^0 - 2e^- \rightarrow Zn^{2+}$$

You may write Zn instead of Zn^0.   [1 point]

55. The salt bridge permits the migration of ions and maintains the neutrality of both electrolytic solutions. Either choice is acceptable as an answer. [1 point]

56. **a** The reaction represents the *fission* of uranium-235.   [1 point]

b The "lost mass" is converted into energy.   [1 point]

c Any one of the following (or similar) responses is acceptable for credit:

- fusion
- nuclear decay
- radioactive decay
- natural transmutation   [1 point]

57. Three possible representations of the NH_3 molecule are shown below. Note that in each representation, nitrogen has one unshared pair of electrons. [1 point]

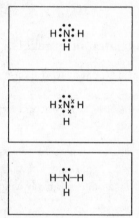

58. Any one of the following (or similar) responses is acceptable for credit:

- The molecule is symmetrical in shape and/or charge.
- The electrons are evenly distributed within the molecule.
- All of the dipoles cancel; the molecule has no dipole moment. [1 point]

59. Any one of the following (or similar) responses is acceptable for credit:

- NH_3 consists of polar molecules that attract each other more strongly than do molecules of Cl_2.
- The nitrogen in NH_3 has an unshared pair of electrons, leading to stronger attractions.
- NH_3 is capable of hydrogen bonding.   [1 point]

60. Any one of the following (or similar) responses is acceptable for credit:

- KCl contains ionic bonds, while Cl_2, CCl_4, and NH_3 contain covalent bonds.
- In KCl, the ions do not share electrons.
- In KCl, the oppositely charged ions attract each other.
- In KCl, the bond is formed between a metal and a nonmetal. [1 point]

61. In unsaturated hydrocarbons, at least one double or triple bond is present between two of the carbon atoms in the molecule. In saturated hydrocarbons, all of the carbon-to-carbon bonds are joined by only single bonds. [1 point]

PART C

[Point values are indicated in brackets.]

62. Either of the following (or similar) responses is acceptable for credit:

- The atom is mostly empty space.
- The volume of the atom is mostly unoccupied. [1 point]

63. Either of the following (or similar) responses is acceptable for credit:

- Alpha particles were deflected by the positively charged nucleus.
- The nucleus of the atom is positively charged. [1 point]

64. Any one of the following (or similar) responses is acceptable for credit:

- The atom has a positively charged nucleus surrounded by negatively charged electrons.
- The positive charges, but not the electrons, are contained in the nucleus.
- The nucleus of the atom is small in comparison with the atom as a whole. [1 point]

65. As the pressure above the cola is decreased, the solubility of the $CO_2(g)$ decreases.

Note: It is *not* sufficient to say that the soda "goes flat."   [1 point]

66. As the temperature of the cola increases, the solubility of the $CO_2(g)$ decreases.   [1 point]

67. **a** The axes must be correctly drawn and labeled. Either set of axes is acceptable for credit: [1 point]

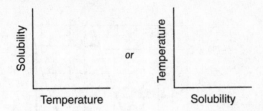

b Any graph that begins at the top left of the graph and continues toward the bottom right is acceptable for credit. Two possibilities are shown below: [1 point]

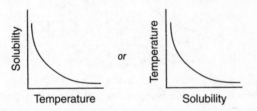

68. Refer to the vapor pressure graph shown below. The vapor pressure of liquid A at 70°C is 710 (± 10) mmHg.

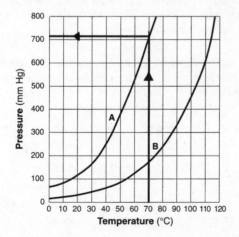

Note: 1 point is awarded for the answer, and 1 point is awarded for the units. [2 points]

69. Refer to the vapor pressure graph shown above. The temperature at which liquid *B* has the same vapor pressure as liquid *A* at 70°C is 114°C (± 2°C).

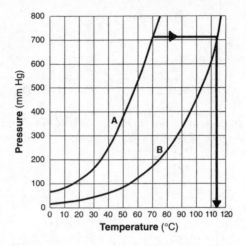

Note: 1 point is awarded for the answer, and 1 point is awarded for the units. [2 points]

70. Liquid *A*. The higher vapor pressure of liquid *A* indicates that the attractions between the molecules of liquid *A* are weaker than those of liquid *B*. These weaker intermolecular forces allow the molecules of liquid *A* to escape into the vapor phase more easily.

Note: 1 point is awarded for identifying the liquid correctly, and one point is awarded for providing an acceptable explanation in terms of intermolecular forces.   [2 points]

71. Subtract the initial NaOH buret reading from the final NaOH buret reading:

$$35.2 \text{ mL} - 23.2 \text{ mL} = \textbf{12.0 mL}$$

Note: 1 point is awarded for setting up the work correctly, and 1 point is awarded for a correct answer.   [2 points]

72. A suitable indicator should change color *at or near* the pH of the end point of the titration. The pH of the end point of this titration (HCl, a strong acid and NaOH, a strong base) is 7. According to Reference Table *M*:

- phenolphthalein is suitable because it changes from colorless to pink after pH = 7;
- litmus reaches an intermediate color approximately at pH = 7 and is therefore suitable;
- bromthymol blue is also suitable because it reaches an intermediate color at approximately pH = 7.

Note: 1 point is awarded for choosing *one* suitable indicator, and 1 point is awarded for providing an acceptable reason for selecting this indicator. [2 points]

73. The molarity of the NaOH can be found from Equation 7 of Reference Table *T*. For example, the molarity of trial 3 is calculated as follows:

$$M_A V_A = M_B V_B$$

$$(1.2 \text{ M}) \cdot (10.0 \text{ mL}) = M_B \cdot (12.0 \text{ mL})$$

$$M_B = \textbf{1.0 M}$$

Refer to the table below in which the molarities for all four trials have been calculated:

	Trial 1	Trial 2	Trial 3	Trial 4
Initial NaOH Buret Reading	0.0 mL	12.2 mL	23.2 mL	35.2 mL
Final NaOH Buret Reading	12.2 mL	23.2 mL	35.2 mL	47.7 mL
Volume of NaOH Used	12.2 mL	11.0 mL	12.0 mL	12.5 mL
Volume of 1.2 M HCl Used	10.0 mL	10.0 mL	10.0 mL	10.0 mL
Molarity of NaOH	0.98 M	1.1 M	1.0 M	0.96 M
Average Molarity	**1.0 M**			

Note: 1 point is awarded for the answer (1, 1.0, or 1.01); 1 point is awarded for expressing the answer to 2 significant figures (1.0); and 1 point is awarded for the correct units (M or mol/L). [3 points]

74. Any one of the following (or similar) responses is acceptable for credit:

- Multiple trials minimize the experimental error within each trial.
- Multiple trials improve the accuracy of the answer.
- The errors in each trial might be above or below the true value. The average of multiple trials contains the least error. [1 point]

Mark (✓) the questions you answered correctly. Count the number of checks and follow the formulas given to determine your score on each topic.

Core Area	☐ Questions Answered Correctly

43, 47, 67a, 67b, 68, 69, 70, 73, 74

Section M—Math Skills
☐ Number of checks ÷ 9 × 100 = _____ %

1, 2, 3, 36, 62, 63, 64

Section I—Atomic Concepts
☐ Number of checks ÷ 7 × 100 = _____ %

6, 7, 9, 37

Section II—Periodic Table
☐ Number of checks ÷ 4 × 100 = _____ %

8, 13, 22, 42, 48

Section III—Moles/Stoichiometry
☐ Number of checks ÷ 5 × 100 = _____ %

6, 10, 13, 14, 57, 58, 59, 60

Section IV—Chemical Bonding
☐ Number of checks ÷ 8 × 100 = _____ %

4, 12, 15, 16, 17, 19, 21, 24, 30, 35, 39, 40, 44, 47,
51a, 51b, 52, 65, 66, 67a, 67b, 68, 69, 70

Section V—Physical Behavior of Matter
☐ Number of checks ÷ 23 × 100 = _____%

5, 11, 38, 41, 45, 50

Section VI—Kinetics and Equilibrium
☐ Number of checks ÷ 6 × 100 = _____ %

18, 25, 26, 49, 61

Section VII—Organic Chemistry
☐ Number of checks ÷ 5 × 100 = _____ %

4, 27, 28, 46, 53, 54, 55

Section VIII—Oxidation-Reduction
☐ Number of checks ÷ 7 × 100 = _____ %

23, 32, 34, 71, 72, 73

Section IX—Acids, Bases, and Salts
☐ Number of checks ÷ 6 × 100 = _____ %

20, 29, 31, 33, 56a, 56b, 56c

Section X—Nuclear Chemistry
☐ Number of checks ÷ 7 × 100 = _____ %

Examination
June 2003

Chemistry
The Physical Setting

PART A

Answer all questions in this part.

Directions (1–35): For *each* statement or question, write in the answer space the *number* of the word or expression that, of those given, best completes the statement or answers the question. Some questions may require the use of the *Reference Tables for Physical Setting/Chemistry.*

1 The atomic number of an atom is always equal to the number of its

(1) protons, only
(2) neutrons, only
(3) protons plus neutrons
(4) protons plus electrons   1_____

2 Which subatomic particle has no charge?

(1) alpha particle   (3) neutron
(2) beta particle   (4) electron   2_____

3 When the electrons of an excited atom return to a lower energy state, the energy emitted can result in the production of

(1) alpha particles   (3) protons
(2) isotopes   (4) spectra   3_____

4 The atomic mass of an element is calculated using the

   (1) atomic number and the ratios of its naturally occurring isotopes
   (2) atomic number and the half-lives of each of its isotopes
   (3) masses and the ratios of its naturally occurring isotopes
   (4) masses and the half-lives of each of its isotopes   4_____

5 The region that is the most probable location of an electron in an atom is

   (1) the nucleus   (3) the excited state
   (2) an orbital   (4) an ion   5_____

6 Which is a property of most nonmetallic solids?

   (1) high thermal conductivity
   (2) high electrical conductivity
   (3) brittleness
   (4) malleability   6_____

7 Alpha particles are emitted during the radioactive decay of

   (1) carbon-14   (3) calcium-37
   (2) neon-19   (4) radon-222   7_____

8 Which is an empirical formula?

   (1) P_2O_5 (3) C_2H_4
   (2) P_4O_6 (4) C_3H_6 8_____

9 Which substance can be decomposed by a chemical change?

   (1) Co   (3) Cr
   (2) CO   (4) Cu   9_____

10 The percent by mass of calcium in the compound calcium sulfate ($CaSO_4$) is approximately

(1) 15%   (3) 34%
(2) 29%   (4) 47%   10_____

11 What is represented by the dots in a Lewis electron-dot diagram of an atom of an element in Period 2 of the Periodic Table?

(1) the number of neutrons in the atom
(2) the number of protons in the atom
(3) the number of valence electrons in the atom
(4) the total number of electrons in the atom   11_____

12 Which type of chemical bond is formed between two atoms of bromine?

(1) metallic   (3) ionic
(2) hydrogen   (4) covalent   12_____

13 Which of these formulas contains the most polar bond?

(1) H–Br   (3) H–F
(2) A–Cl   (4) H–I   13_____

14 According to Table *F*, which of these salts is *least* soluble in water?

(1) LiCl   (3) $FeCl_2$
(2) RbCl   (4) $PbCl_2$ 14_____

15 Which of these terms refers to matter that could be heterogeneous?

(1) element   (3) compound
(2) mixture   (4) solution   15_____

16 In which material are the particles arranged in a regular geometric pattern?

(1) $CO_2(g)$ (3) $H_2O(\ell)$

(2) $NaCl(aq)$ (4) $C_{12}H_{22}O_{11}(s)$ 16_____

17 Which change is exothermic?

(1) freezing of water

(2) melting of iron

(3) vaporization of ethanol

(4) sublimation of iodine   17_____

18 Which type of change must occur to form a compound?

(1) chemical   (3) nuclear

(2) physical   (4) phase   18_____

19 Which formula correctly represents the composition of iron (III) oxide?

(1) FeO_3 (3) Fe_3O

(2) Fe_2O_3 (4) Fe_3O_2 19_____

20 Given the reaction:

$$PbCl_2(aq) + Na_2CrO_4(aq) \rightarrow$$
$$PbCrO_4(s) + 2\ NaCl(aq)$$

What is the total number of moles of NaCl formed when 2 moles of Na_2CrO_4 react completely?

(1) 1 mole   (3) 3 moles

(2) 2 moles   (4) 4 moles   20_____

21 Which hydrocarbon is saturated?

(1) propene   (3) butene

(2) ethyne   (4) heptane   21_____

22 Which statement correctly describes an endothermic chemical reaction?

   (1) The products have higher potential energy than the reactants, and the ΔH is negative.
   (2) The products have higher potential energy than the reactants, and the ΔH is positive.
   (3) The products have lower potential energy than the reactants, and the ΔH is negative.
   (4) The products have lower potential energy than the reactants, and the ΔH is positive.   22_____

23 At standard pressure when NaCl is added to water, the solution will have a

   (1) higher freezing point and a lower boiling point than water
   (2) higher freezing point and a higher boiling point than water
   (3) lower freezing point and higher boiling point than water
   (4) lower freezing point and a lower boiling point than water   23_____

24 Which element has atoms that can form single, double, and triple covalent bonds with other atoms of the same element?

   (1) hydrogen   (3) fluorine
   (2) oxygen   (4) carbon   24_____

25 Which compound is an isomer of pentane?

   (1) butane   (3) methyl butane
   (2) propane   (4) methyl propane   25_____

26 In which substance does chlorine have an oxidation number of +1?

(1) Cl_2 (3) $HClO$

(2) HCl (4) $HClO_2$ 26____

27 Which statement is true for any electrochemical cell?

(1) Oxidation occurs at the anode, only.

(2) Reduction occurs at the anode, only.

(3) Oxidation occurs at both the anode and the cathode.

(4) Reduction occurs at both the anode and the cathode.   27____

28 Given the equation:

$$2\ Al + 3\ Cu^{2+} \rightarrow 2\ AL^{3+} + 3\ Cu$$

The reduction half-reaction is

(1) $Al \rightarrow Al^{3+} + 3e^-$ (3) $Al + 3e^- \rightarrow Al^{3+}$

(2) $Cu^{2+} + 2e^- \rightarrow Cu$ (4) $Cu^{2+} \rightarrow Cu + 2e^-$ 28____

29 Which 0.1 M solution contains an electrolyte?

(1) $C_6H_{12}O_6(aq)$ (3) $CH_3OH(aq)$

(2) $CH_3COOH(aq)$ (4) $CH_3OCH_3(aq)$ 29____

30 Which equation represents a neutralization reaction?

(1) $Na_2CO_3 + CaCl_2 \rightarrow 2\ NaCl + CaCO_3$

(2) $Ni(NO_3)_2 + H_2S \rightarrow NiS + 2\ HNO_3$

(3) $NaCl + AgNO_3 \rightarrow AgCl + NaNO_3$

(4) $H_2SO_4 + Mg(OH)_2 \rightarrow MgSO_4 + 2\ H_2O$ 30____

31 An Arrhenius acid has

(1) only hydroxide ions in solution
(2) only hydrogen ions in solution
(3) hydrogen ions as the only positive ions in solution
(4) hydrogen ions as the only negative ions in solution　　　31_____

32 Which type of radioactive emission has a positive charge and weak penetrating power?

(1) alpha particle　　　(3) gamma ray
(2) beta particle　　　(4) neutron　　　32_____

33 Which substance contains metallic bonds?

(1) $Hg(\ell)$　　　(3) $NaCl(s)$
(2) $H_2O(\ell)$　　　(4) $C_6H_{12}O_6(s)$　　　33_____

34 What is the name of the process in which the nucleus of an atom of one element is changed into the nucleus of an atom of a different element?

(1) decomposition　　　(3) substitution
(2) transmutation　　　(4) reduction　　　34_____

Note that question 35 has only three choices.

35 A catalyst is added to a system at equilibrium. If the temperature remains constant, the activation energy of the forward reaction

(1) decreases
(2) increases
(3) remains the same　　　35_____

PART B–1

Answer all questions in this part.

Direction (36–50): For *each* statement or question, write in the answer space the *number* of the word or expression that, of those given, best completes the statement or answers the question. Some questions may require the use of the *Reference Tables for Physical Setting/Chemistry*.

36 The nucleus of an atom of K-42 contains

    (1) 19 protons and 23 neutrons
    (2) 19 protons and 42 neutrons
    (3) 20 protons and 19 neutrons
    (4) 23 protons and 19 neutrons   36_____

37 What is the total number of electrons in a CU^+ ion?

    (1) 28   (3) 30
    (2) 29   (4) 36   37_____

38 Which list of elements is arranged in order of increasing atomic radii?

    (1) Li, Be, B, C   (3) Sc, Ti, V, Cr
    (2) Sr, Ca, Mg, Be   (4) F, Cl, Br, I   38_____

39 Which isotope is most commonly used in the radioactive dating of the remains of organic materials?

    (1) ^{14}C (3) ^{32}P
    (2) ^{16}N (4) ^{37}K 39_____

40 According to Reference Table *H*, what is the vapor pressure of propanone at 45°C?

    (1) 22 kPa   (3) 70. kPa
    (2) 33 kPa   (4) 98 kPa   40_____

41 The freezing point of bromine is

(1) 539°C   (3) 7°C

(2) –539°C   (4) –7°C   41_____

42 Hexane (C_6H_{14}) and water do *not* form a solution. Which statement explains this phenomenon?

(1) Hexane is polar and water is nonpolar.

(2) Hexane is ionic and water is polar.

(3) Hexane is nonpolar and water is polar.

(4) Hexane is nonpolar and water is ionic.   42_____

43 The potential energy diagram below represents a reaction.

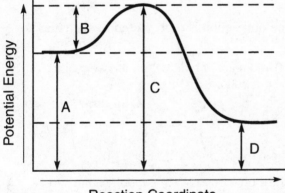

Reaction Coordinate

Which arrow represents the activation energy of the forward reaction?

(1) A   (3) C

(2) B   (4) D   43_____

44 Given the formulas of four organic compounds:

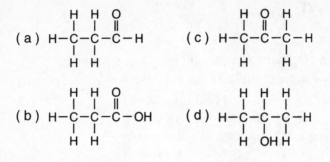

Which pair below contains an alcohol and an acid?

(1) *a* and *b* (3) *b* and *d*
(2) *a* and *c* (4) *c* and *d* 44_____

45 Which type of reaction is represented by the equation below?

Note: n and n are very large numbers equal to about 2000.

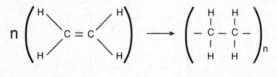

(1) esterification (3) saponification
(2) fermentation (4) polymerization 45_____

46 A diagram of a chemical cell and an equation are shown below.

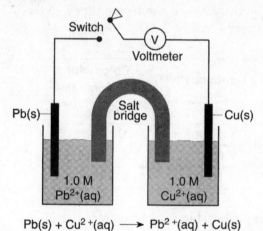

$$Pb(s) + Cu^{2+}(aq) \longrightarrow Pb^{2+}(aq) + Cu(s)$$

When the switch is closed, electrons will flow from

(1) the $Pb(s)$ to the $Cu(s)$
(2) the $Cu(s)$ to the $Pb(s)$
(3) the $Pb^{2+}(aq)$ to the $Pb(s)$
(4) the $CU^{2+}(aq)$ to the $Cu(s)$

46_____

47 Which ion has the same electron configuration as an atom of HE?

(1) H^- (3) Na^+
(2) O^{2-} (4) Ca^{2+}

47_____

48 A student was given four unknown solutions. Each solution was checked for conductivity and tested with phenolphthalein. The results are shown in the data table below.

Solution	Conductivity	Color with Phenolphthalein
A	Good	Colorless
B	Poor	Colorless
C	Good	Pink
D	Poor	Pink

Based on the data table, which unknown solution could be 0.1 M NaOH?

(1) A   (3) C

(2) B   (4) D   48_____

49 In the reaction $^{239}_{93}Np \rightarrow {}^{239}_{94}Pu + X$, what does X represent?

(1) a neutron   (3) an alpha particle

(2) a proton   (4) a beta particle   49_____

Note that question 50 has only three choices.

50 As carbon dioxide sublimes, its entropy

(1) decreases

(2) increases

(3) remains the same   50_____

PART B–2

Answer all questions in the part.

Directions (51–63): Record your answers on the answer sheet provided in the back. Some questions may require the use of the *Reference Tables for Physical Setting/Chemistry.*

Base your answers to questions 51 and 52 on the electron configuration table shown below:

Element	Electron Configuration
X	2–8–8–2
Y	2–8–7–3
Z	2–8–8

51 What is the total number of valence electrons in an atom of electron configuration *X*?   [1]

52 Which electron configuration represents the excited state of a calcium atom?   [1]

Base your answers to questions 53 and 54 on the information below.

Given: Samples of Na, Ar, As, Rb

53 Which *two* of the given elements have the most similar chemical properties?   [1]

54 Explain your answer in terms of the Periodic Table of the Elements.   [1]

Base your answers to questions 55 and 56 on the information below.

Diethyl ether is widely used as a solvent.

55 In the space provided *in the back,* draw the structural formula for diethyl ether.   [1]

56 In the space provided *in the back,* draw the structural formula for an alcohol that is an isomer of diethyl ether.   [1]

Base your answers to questions 57 and 58 on the information below.

Two chemistry students each combine a different metal with hydrochloric acid. Student A uses zinc, and hydrogen gas is readily produced. Student *B* uses copper, and no hydrogen gas is produced.

57 State one chemical reason for the different results of students *A* and *B.*   [1]

58 Using Reference Table *J,* identify another metal that will react with hydrochloric acid to yield hydrogen gas.   [1]

59 Given the reaction between two different elements in the gaseous state:

Box *A* below represents a mixture of the two reactants before the reaction occurs. The product of this reaction is a gas. In Box *B* provided *in the back,*

draw the system after the reaction has gone to completion, based on the Law of Conservation of Matter. [2]

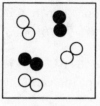

Box A
System Before Reaction

60 As a neutral sulfur atom gains two electrons, what happens to the radius of the atom? [1]

61 After a neutral sulfur atom gains two electrons, what is the resulting charge of the ion? [1]

62 *a* In the space provided *in the back,* calculate the heat released when 25.0 grams of water freezes at 0°C. Show all work. [1]

b Record your answer with an appropriate unit. [1]

63 State one difference between voltaic cells and electrolytic cells. Include information about *both* types of cells in your answer. [1]

PART C

Answer all questions in this part.

Directions (64–79): Record your answers on the answer sheet provided in the back. Some questions may require the use of the *Reference Tables for Physical Setting/Chemistry*.

Base your answers to questions 64 and 65 on the diagram below, which shows a piston confining a gas in a cylinder.

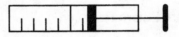

64 Using the set of axes provided *in the back*, sketch the general relationship between the pressure and the volume of an ideal gas at constant temperature. [1]

65 The gas volume in the cylinder is 6.2 milliliters and its pressure is 1.4 atmospheres. The piston is then pushed in until the gas volume is 3.1 milliliters while the temperature remains constant.

   a In the space provided *in the back,* calculate the pressure, in atmospheres, after the change in volume. Show all work. [1]
   b Record your answer. [1]

66 A student recorded the following buret readings during a titration of a base with an acid:

	Standard 0.100 M HCl	Unknown KOH
Initial reading	9.08 mL	0.55 mL
Final reading	19.09 mL	5.56 mL

   a In the space provided *in the back,* calculate the molarity of the KOH. Show all work.   [1]

   b Record your answer to the correct number of significant figures.   [1]

67 John Dalton was an English scientist who proposed that atoms were hard, indivisible spheres. In the modern model, the atom has a different internal structure.

   a Identify one experiment that led scientists to develop the modern model of the atom.   [1]

   b Describe this experiment.   [1]

   c State one conclusion about the internal structure of the atom, based on this experiment.   [1]

Base your answers to questions 68 through 73 on the information below and on your knowledge of chemistry.

Nuclear Waste Storage Plan for Yucca Mountain

   In 1978, the U.S. Department of Energy began a study of Yucca Mountain which is located 90 miles from Las Vegas, Nevada. The study was to determine if Yucca Mountain would be suitable for a long-term burial site for high-level radioactive waste. A three-dimensional (3-D) computer scale model of the site was used to simulate the Yucca Mountain area. The computer model study for Yucca Mountain included such variables as: the possibility of earthquakes, predicted water flow through the mountain, increased rainfall due to climate changes, radioactive leakage from the waste containers, and increased temperatures from the buried waste within the containers.

The containers that will be used to store the radioactive waste are designed to last 10,000 years. Within the 10,000-year time period, cesium and strontium, the most powerful radioactive emitters, would have decayed. Other isotopes found in the waste would decay more slowly, but are not powerful radioactive emitters.

In 1998, scientists discovered that the compressed volcanic ash making up Yucca Mountain was full of cracks. Because of the arid climate, scientists assumed that rainwater would move through the cracks at a slow rate. However, when radioactive chlorine-36 was found in rock samples at levels halfway through the mountain, it was clear that rainwater had moved quickly down through Yucca Mountain. It was only 50 years earlier when this chlorine-36 isotope had contaminated rainwater during atmospheric testing of the atom bomb.

Some opponents of the Yucca Mountain plan believe that the uncertainties related to the many variables of the computer model result in limited reliability of its predictions. However, advocates of the plan believe it is safer to replace the numerous existing radioactive burial sites around the United States with the one site at Yucca Mountain. Other opponents of the plan believe that transporting the radioactive waste to Yucca Mountain from the existing 131 burial sites creates too much danger to the United States. In 2002, after many years of political debate, a final legislative vote approved the development of Yucca Mountain to replace the existing 131 burial sites.

68 State one uncertainty in the computer model that limits the reliability of this computer model. [1]

69 Scientists assume that a manufacturing defect would cause at least one of the waste containers stored in the Yucca Mountain repository to leak within the first 1,000 years. State one possible effect such a leak could have on the environment near Yucca Mountain. [1]

70 State one risk associated with leaving radioactive waste in the 131 sites around the country where it is presently stored. [1]

71 If a sample of cesium-137 is stored in a waste container in Yucca Mountain, how much time must elapse until only $\frac{1}{32}$ of the original sample remains unchanged? [1]

72 The information states "Within the 10,000-year time period, cesium and strontium, the most powerful radioactive emitters, would have decayed." Use information from Reference Table N to support this statement. [1]

73 Why is water flow a crucial factor in deciding whether Yucca Mountain is a suitable burial site? [1]

Base your answers to questions 74 through 76 on the information below.

A student wishes to investigate how the reaction rate changes with a change in concentration of $HCl(aq)$.

Given the reaction: $Zn(s) + HCl(aq) \rightarrow$ $H_2(g) + ZnCl_2(aq)$

74 Identify the independent variable in this investigation.   [1]

75 Identify one other variable that might affect the rate and should be held constant during this investigation.   [1]

76 Describe the effect of increasing the concentration of $HCl(aq)$ on the reaction rate and justify your response in terms of *collision theory*.   [1]

Base your answers to questions 77 through 79 on the information below.

A truck carrying concentrated nitric acid overturns and spills its contents. The acid drains into a nearby pond. The pH of the pond water was 8.0 before the spill. After the spill, the pond water is 1,000 times more acidic.

77 Name an ion in the pond water that has increased in concentration due to this spill.   [1]

78 What is the new pH of the pond water after the spill?   [1]

79 What color would bromthymol blue be at this new pH?   [1]

Answer Sheet
June 2003

Chemistry
The Physical Setting

PART B–2

Answer Space

51 _____

52 _____

53 _____ and _____

54 _____

_____

55

56

57 _____

_____

58 _____

59

Box A
System Before Reaction

Box B
System After Reaction Has
Gone to Completion

60 _____

_____

61 _____

62 *a*

b _____

63 _____

_____

_____

PART C

Answer Space

64

65 *a*

  b _____ **atm**

66 *a*

  b _____ **M**

67 *a* _____
  b _____
   _____
   _____
  c _____
   _____
   _____

68 _____

69 _____

70 _____

71 _____

72 _____

73 _____

74 _____

75 _____

76 _____

77 _____

78 _____

79 _____

Answers
June 2003

Chemistry
The Physical Setting

Answer Key

PART A

1. 1	8. 1	15. 2	22. 2	29. 2
2. 3	9. 2	16. 4	23. 3	30. 4
3. 4	10. 2	17. 1	24. 4	31. 3
4. 3	11. 3	18. 1	25. 3	32. 1
5. 2	12. 4	19. 2	26. 3	33. 1
6. 3	13. 3	20. 4	27. 1	34. 2
7. 4	14. 4	21. 4	28. 2	35. 1

PART B–1

36. 1	39. 1	42. 3	45. 4	48. 3
37. 1	40. 3	43. 2	46. 1	49. 4
38. 4	41. 4	44. 3	47. 1	50. 2

Answers Explained

PART A

1. **1** The atomic number of an atom is defined as the nuclear charge, that is, the number of protons in the nucleus.

Wrong Choice Explained:
(3) The number of protons plus neutrons in the nucleus is known as the *mass number*.

2. **3** The term *neutron* means an electrically neutral particle.

Wrong Choices Explained:
(1) An alpha particle is a helium-4 nucleus, which is positively charged.
(2), (4) A beta particle is an electron, which is negatively charged.

3. **4** As an electron in an excited atom returns to a lower energy level, it emits a photon of light. Collectively, all of the emitted photons are known as *spectra*.

4. **3** In order to calculate the (average) atomic mass of an element, one must know which isotopes occur naturally, the masses of each of these isotopes, and the abundance of these isotopes.

5. **2** An orbital is defined as the region in which an electron in an atom is most probably located.

6. **3** Nonmetallic solids do not contain mobile valence electrons. As a result, they do not conduct heat or electricity well and they are not easily bent into shape (that is, malleable). The rigidity of their crystal structure makes them brittle solids.

7. **4** Refer to Reference Table N. Of the choices given, only choice (4), radon-222 (^{222}Rn), is an alpha particle emitter.

8. **1** An empirical formula is one in which the elements appear in smallest whole number ratios. Of the choices given, only choice (1), P_2O_5, is an empirical formula.

Wrong Choices Explained:
(2) The empirical formula of this compound is P_2O_3.
(3), (4) The empirical formula of each of these compounds is CH_2.

9. **2** Only compounds can be decomposed by a chemical change. Of the choices given, only choice (2), CO (carbon monoxide), is a compound.

Wrong Choices Explained:
(1), (3), (4) Co (cobalt), Cr (chromium), and Cu (copper) are elements and cannot be decomposed by a chemical change.

10. **2** First, we complete the following table for $CaSO_4$:

Element	Atomic mass (g/mol)	Number of Atoms in Formula	Mass of Element in Formula/g
Ca	40	1	40
S	32	1	32
O	16	4	64
		Formula mass	**136**

Since the contribution of oxygen to the formula mass is 64, we can calculate the percent composition by mass from the following relationship:

$$\frac{40}{136} \times 100 = \mathbf{29\%}$$

11. **3** In Period 2, a Lewis electron-dot diagram is a representation of the filling pattern of the valence electrons in a particular atom. Each dot represents a particular valence electron.

12. **4** Use the Periodic Table of the Elements. Bromine (Br, element number 35) has seven valence electrons. When two atoms of bromine combine, a single pair of electrons is shared equally between the atoms. This is known as a (nonpolar) covalent bond.

13. **3** Use the electronegativity values found on Reference Table S. The bond with the greatest electronegativity difference will be the most polar bond. Choice (3), H–F, has an electronegativity difference of 1.9 (4.0 − 2.1); it is the largest electronegativity difference of the four choices given in this question.

14. **4** Note that all of the compounds listed in this question are chlorides. Refer to Reference Table *F*. Halides—including chlorides (Cl⁻)—are generally soluble in water, with the principal exceptions of Ag^+, Pb^{2+}, and Hg_2^{2+}. Choice (4) lists a chloride containing Pb^{2+}.

15. **2** The term *heterogeneous* means nonuniform, such as an ice cream soda. Only mixtures can be heterogeneous.

Wrong Choice Explained:
  (4) A solution is a mixture, but it is a uniform mixture.

16. **4** When the particles of a substance are arranged in a regular geometric pattern, the result is a crystalline solid. Of the choices given, only choice (4), $C_{12}H_{22}O_{11}(s)$, represents a substance in the solid phase.

17. **1** The term *exothermic* means that energy is released as a process occurs. Refer to the table below, which shows the energy changes accompanying phase changes:

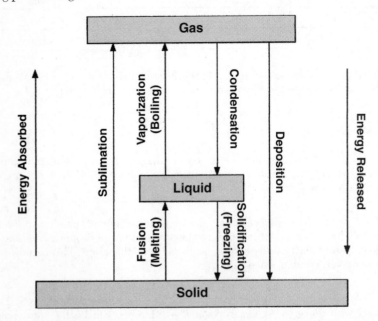

Of the choices given, only choice (1), freezing of water, is accompanied by the release of energy.

18. **1** When a compound is produced, chemical bonds are broken and other chemical bonds are formed. This production of a new substance is the main characteristic of a chemical change.

19. **2** The oxide ion has a charge of 2^-, while the iron (III) ion has a charge of $3+$. In the compound, the sum of the charges must total zero. Of the choices given, only choice (2), Fe_2O_3, meets this requirement.

20. **4** According to the balanced equation, 2 moles of NaCl will be formed when 1 mole of Na_2CrO_4 reacts completely. If the quantity of Na_2CrO_4 that reacts is doubled (2 moles), twice as much NaCl (4 moles) will be formed.

21. **4** A saturated hydrocarbon is one that contains only single carbon-to-carbon bonds. Refer to Reference Table Q. Names of saturated hydrocarbons end in –ane, such as choice (4), heptane.

22. **2** The term *endothermic* means that energy is absorbed as the result of a chemical reaction. As a result, the products have a higher potential energy than the reactants. Since ΔH represents the difference in potential energy ($\Delta H = H_{products} - H_{reactants}$), the sign of ΔH will be positive.

23. **3** When a nonvolatile solute, such as NaCl, is added to water, the vapor pressure of the solution is lowered. This change results in the solution having a lower freezing point and a higher boiling point than pure water.

24. **4** Of the choices given, only choice (4), carbon, can form single, double, or triple bonds with another carbon atom.

Wrong Choices Explained:
(1), (3) Two atoms of hydrogen or two atoms of fluorine can form only single covalent bonds between themselves.
(2) Two atoms of oxygen can form either single or double covalent bonds between themselves.

25. **3** Isomers have the same molecular formulas but different structural formulas. The structural and molecular formulas of each of the compounds named in this question are given below:

pentane

```
      H   H   H   H   H
      |   |   |   |   |
 H — C — C — C — C — C — H    C₅H₁₂
      |   |   |   |   |
      H   H   H   H   H
```

C_5H_{12}

(1) butane

```
      H   H   H   H
      |   |   |   |
 H — C — C — C — C — H        C₄H₁₀
      |   |   |   |
      H   H   H   H
```

C_4H_{10}

(2) propane

```
      H   H   H
      |   |   |
 H — C — C — C — H            C₃H₈
      |   |   |
      H   H   H
```

C_3H_8

(3) methyl butane

```
          H
          |
      H H-C-H H      H
      |   |   |      |
 H — C — C — C — C — H        C₅H₁₂
      |   |   |      |
      H   H   H      H
```

C_5H_{12}

(4) methyl propane

```
          H
          |
      H H-C-H H
      |   |   |
 H — C — C — C — H            C₄H₁₀
      |   |   |
      H   H   H
```

C_4H_{10}

Of the choices given, only choice (3), methyl butane, has the same molecular formula and a different structural formula than pentane.

26. **3** In all compounds, the sum of the oxidation numbers must add to 0. In choices (2), (3), and (4), given in this question, hydrogen has an oxidation number of +1 and each oxygen (O) has an oxidation number of –2. In choice (3), HClO, the H accounts for +1 and the O atom accounts for –2. For the oxidation numbers to add to 0, chlorine (Cl) must have an oxidation number of +1.

Wrong Choices Explained:
(1) In Cl_2, the oxidation number of Cl is 0.
(2) In HCl, the oxidation number of Cl is –1.
(4) In $HClO_2$, the oxidation number of Cl is +3.

27. **1** The term *anode* is defined as the electrode at which oxidation occurs. This is true for all electrochemical cells—voltaic and electrolytic.

28. **2** When a particle undergoes reduction, it gains electrons and its oxidation number decreases. The half-reaction given in choice (2) shows the gain of 2 electrons by Cu^{2+} as its oxidation number changes from +2 to 0.

Wrong Choice Explained:
(1) This half-reaction correctly represents the oxidation of Al.
(3), (4) Both of these are incorrectly written half-reactions.

29. **2** An electrolyte is a substance that forms ions in solution and, as a result, the solution conducts electricity. Acetic acid, CH_3COOH, is a weak electrolyte that ionizes according to the equation:

$$CH_3COOH + H_2O \rightleftharpoons H_3O^+ + CH_3COO^-$$

Wrong Choices Explained:
(1), (3), (4) These molecular substances, glucose, methanol, and dimethyl ether, do not form ions in aqueous solution.

30. **4** In a neutralization reaction, an acid and a hydroxide base combine to form a salt and water. In choice (4), $H_2SO_4 + Mg(OH)_2 \rightarrow MgSO_4 + 2H_2O$, sulfuric acid reacts with magnesium hydroxide to form magnesium sulfate and water.

Wrong Choices Explained:
(1), (2), (3) In none of these reactions does an acid react with a base.

31. **3** An Arrhenius acid is defined as a substance that produces hydrogen ions as the only positive ions in solution.

32. **1** An alpha particle is the nucleus of a helium-4 atom. Of the choices given in the question, it is the only particle that has a positive charge. In addition, its penetrating power is the weakest of the four particles.

Wrong Choices Explained:
(2) A beta particle is an electron and has a negative charge. Its penetrating power is weak but not as weak as that of an alpha particle.
(3) A gamma ray has no charge and has an extremely strong penetrating power.
(4) A neutron has no charge and has a moderate penetrating power.

33. **1** Metallic bonds are found in metallic substances. The presence of metallic bonds confers good conductivity, malleability, and luster on the substance. Of the choices given, only choice (1), $Hg(\ell)$, is a metallic substance. The phase of the mercury (liquid) is not important in answering this question.

Wrong Choices Explained:
(2), (4) H_2O and $C_6H_{12}O_6$ are covalently bonded compounds and do not possess metallic properties.
(3) NaCl is an ionic compound and does not possess metallic properties.

34. **2** Transmutation is defined as the change that occurs when one nucleus is transformed into another.

Wrong Choices Explained:
(1) Decomposition is a chemical reaction in which a compound is changed into simpler substances.
(3) Substitution is a reaction in which a hydrogen atom in an organic compound is replaced by the atom of another element.
(4) Reduction is a half-reaction in which electrons are gained by a substance.

35. **1** Catalysts function by decreasing the activation energy of a reaction.

PART B–1

36. **1** The symbol K-42 stands for the isotope of potassium whose mass number is 42. Acording to the Periodic Table of the Elements, the atomic number of potassium is 19. Therefore, the nucleus of an atom of this element contains 19 protons. If the atomic number is subtracted from the mass number of the isotope, the result yields the number of neutrons in the nucleus. (42 protons plus neutrons – 19 protons = 23 neutrons)

37. **1** Acording to the Periodic Table of the Elements, the atomic number of Cu is 29. Therefore, a neutral atom of Cu contains 29 protons and 29 electrons. A Cu^+ ion contains one less electron than an atom of Cu, or 28 electrons.

38. **4** The atomic radius within a group of elements increases as the atomic number increases. The elements of choice (4), F, Cl, Br, I, are all in Group 17 and are listed in order of increasing atomic number. To verify this trend, refer to Reference Table S.

39. **1** In theory, ^{14}C, ^{16}N, or ^{32}P could all be used because organic materials contain these three elements. However, according to Reference Table N, the half-lives of ^{16}N and ^{32}P are much too short (7.2 seconds and 14.2 days, respectively) to be used for radioactive dating. ^{14}C, however, has a half-life of 5730 years and, therefore, is far more suitable for the dating process.

40. **3** Refer to the diagram below:

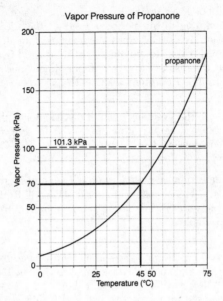

Vapor Pressure of Propanone

The diagram is taken from Reference Table *H*. The bold lines locate the temperature of 45°C and the corresponding vapor pressure of 70 kPa.

41 **4** Use Reference Table *S* and locate bromine (element number 35). The melting point of bromine, which is also its freezing point, is given as 266K. Now use Equation 9 on Reference Table *T* to convert this temperature to the Celsius scale:

$$K = °C + 273$$
$$°C = K - 273$$
$$= 266 - 273$$
$$= \mathbf{-7°C}$$

42. **3** In order for two substances to be able to form a solution, the intermolecular forces must be similar in type and strength. Hexane is a symmetrical nonpolar molecule, and the intermolecular forces (known as *London forces*) are weak. Water is a nonsymmetrical polar molecule, and the intermolecular forces (known as *dipole attractions*) are strong. The differences in the

strengths of London forces and dipole attractions do not permit a solution of hexane and water to form.

43. **2** Refer to the diagram below:

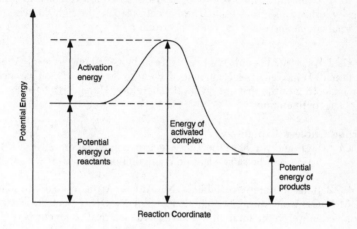

Arrow *B* is the activation energy of the forward reaction.

Wrong Choices Explained:
(1) Arrow *A* is the potential energy of the reactants.
(3) Arrow *C* is the potential energy of the activated complex.
(4) Arrow *D* is the potential energy of the products.

44. **3** Refer to Reference Table *R*. Choice (a) is an *aldehyde*, known as propanal. Choice (b) is an *organic acid*, known as propanoic acid. Choice (c) is a *ketone*, known as propanone. Choice (d) is an *alcohol*, known as 2–propanol. Therefore, pair *b* and *d* contains an alcohol and an acid.

45. **4** The equation shows the linking together of a large number of small molecules (in this case, ethene) to form a single molecule that is very large in size. This process is known as polymerization.

Wrong Choices Explained:
(1) Esterification is the reaction between an organic acid and an alcohol to produce water and an ester.

(2) Fermentation is the anaerobic oxidation of a sugar molecule, such as glucose, to form ethanol and carbon dioxide.

(3) Saponification is the hydrolysis of a lipid using a base such as NaOH or KOH. As a result of this reaction, glycerol and a soap are formed.

46. 1 When the switch is closed, atoms of Pb(s) are oxidized to Pb^{2+}(aq) ions by losing 2 electrons each. These electrons pass through the wire in the external circuit and enter the Cu(s) electrode.

47. 1 An atom of helium has two electrons in its first energy level. Although an atom of H has only 1 electron in its first energy level, an H^- ion has an additional electron in the first energy level. Therefore, He and H^- have the same electron configurations.

Wrong Choices Explained:
(2), (3) O^{2-} and Na^+ have the same electron configuration as Ne (2–8).
(4) Ca^{2+} has the same electron configuration as Ar (2–8–8).

48. 3 Refer to Reference Table M. NaOH is an Arrhenius base. In solution, 0.1 M NaOH would turn phenolphthalein pink. NaOH is an electrolyte; when dissolved in water, the solution will have good electrical conductivity.

49. 4 To balance a nuclear equation, you must be certain that the atomic numbers (the subscripts) and the mass numbers (the superscripts) are equal on both sides of the equation. Apply this rule to the given equation to obtain:

$$^{239}_{93}\text{Np} \rightarrow\ ^{239}_{94}\text{Pu} +\ ^{0}_{-1}X$$

By comparing X with the particles listed on Reference Table O, we see that X is a beta particle.

50. 2 Sublimation is the direct change of a substance from the solid to the gaseous state. As the carbon dioxide sublimes, the molecules become more disordered and the entropy of the system increases.

PART B–2

[Point values are indicated in brackets.]

51. The term *valence electron* refers to an electron in the outermost level of an atom. Since the energy level configuration of X is 2–8–8–2, this atom has 2 valence electrons.   [1 point]

52. Refer to the Periodic Table of the Elements. An atom of calcium (Ca) has 20 electrons and a ground state electron configuration of 2–8–8–2. In an excited state, a calcium atom would still have 20 electrons, but one or more of these electrons would be promoted to a higher energy level. Element Y, whose electron configuration is 2–8–7–3, is an atom of calcium (it has 20 electrons). However, one of the electrons in the third energy level has been promoted to the fourth energy level.   [1 point]

53. Na and Rb would have the most similar chemical properties. (See question 54 for a detailed explanation.)   [1 point]

54. Refer to the Periodic Table of the Elements. Elements that are located within the same group have similar chemical properties. Na and Rb are both located in Group 1 of the Periodic Table of the Elements.   [1 point]

55. Refer to Reference Table R. Diethyl ether contains two ethyl groups on either side of an oxygen atom. Acceptable responses include:

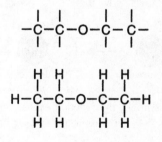

[1 point]

56. Two compounds are isomers if they have the same molecular formula but different structural formulas. Acceptable responses include the following:

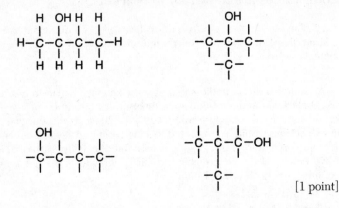

[1 point]

57. Refer to Reference Table J. Any metal that appears above H_2 on the activity series will liberate H_2 gas when reacted with an acid such as HCl. Zinc (Zn) lies above H_2 on this table; it will produce H_2 gas. Copper (Cu) lies below H_2; it will not produce H_2 gas.   [1 point]

58. Choose any metal that lies above H_2 on the table, such as aluminum (Al) or magnesium (Mg).   [1 point]

59. Refer to the diagram below, in which the particle diagram has been rewritten with chemical symbols above it:

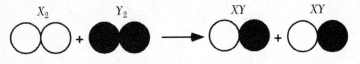

When using chemical symbols, the reaction can be written as $X_2 + Y_2 \rightarrow 2XY$. Note that 1 molecule of X_2 and 1 molecule of Y_2 combine to form 2 molecules of XY.

In Box A, we find 3 molecules of X_2 and 2 molecules of Y_2. Since the combining ratio of X_2 and Y_2 is 1:1, we can conclude that only 2 molecules of X_2 will combine with the 2 molecules of Y_2 and form 4 molecules of XY. 1 molecule of

X_2 will remain unreacted. Therefore, after the reaction has gone to completion, the contents of box B will appear as shown below:

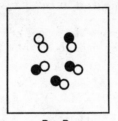

Box B
System After Reaction Has
Gone to Completion

The Law of Conservation of Matter has been obeyed: 6 atoms of X and 4 atoms of Y are present both before the reaction and after the completion of the reaction.

Note that one point is awarded for drawing 4 molecules of XY and 2 molecules of X_2, and one point is awarded for drawing the particles in the gaseous state, that is, randomly arranged.   [2 points]

60. When an atom gains electrons to form a negative ion, its radius increases in size.   [1 point]

61. Each electron has a charge of 1–. When the atom gains 2 electrons, its charge changes from 0 to 2–.   [1 point]

62. Refer to Reference Table B. For every 1 gram of water that freezes, 334 joules of heat are released. Use Equation 8 on Reference Table T:

$$q = mH_f$$
$$= (25.0 \text{ g}) \bullet (334 \text{ J/g})$$
$$= \textbf{8350 J}$$

Note that one point is awarded for setting up the equation correctly, and one point is awarded for displaying the correct answer, including units.   [2 points]

63. A voltaic cell involves a spontaneous redox reaction and, as a result, electrical energy is produced. An electrolytic cell consumes electrical energy in order to force a nonspontaneous redox reaction to occur.   [1 point]

PART C

[Point values are indicated in brackets.]

64. At constant temperature, the volume of a gas is *inversely proportional* to the pressure. Refer to the graphs below. Any one of these graphs is acceptable for credit:

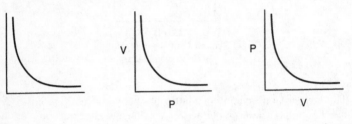

[1 point]

65. Use Equation 6 on Reference Table *T*. At constant temperature, the temperature variable need not be included in the equation:

$$P_1V_1 = P_2V_2$$

$$(1.4 \text{ atm}) \bullet (6.2 \text{ mL}) = P_2 \bullet (3.1 \text{ mL})$$

$$P_2 = \frac{(1.4 \text{ atm}) \bullet (6.2 \text{ mL})}{(3.1 \text{ mL})}$$

$$= \textbf{2.8 atm}$$

Note that one point is award for setting up the equation correctly, and one point is awarded for displaying the correct answer.   [2 points]

66. • The volume of the HCl used is 19.09 mL – 9.08 mL = 10.01 mL.

    • The volume of the NaOH used is 5.56 mL – 0.55 mL = 5.01 mL.

Now use Equation 7 on Reference Table T:

$$M_A V_A = M_B V_B$$

$$(0.100 \text{ M}) \bullet (10.01 \text{ mL}) = M_B \bullet (5.01 \text{ mL})$$

$$M_B = \frac{(0.100 \text{ M}) \bullet (10.01 \text{ mL})}{(5.01 \text{ mL})}$$

$$= \textbf{0.200 M}$$

Note that one point is award for setting up the equation correctly, and one point is awarded for displaying the correct answer expressed to *three* significant figures. [2 points]

67. **a** Of the many experiments that led to the development of the modern model of the atom, perhaps the most familiar one is Rutherford's bombardment of gold foil with alpha particles. (Other experiments include Thomson's work with cathode rays and Moseley's work with X-ray spectra.) [1 point]

b In the gold foil experiment, Rutherford studied the deflection patterns of the alpha particles. [1 point]

c As a result of his studies, Rutherford concluded that the atom was mostly empty space and that most of the mass of the atom was concentrated in a small, positively charged core. [1 point]

68. In order to answer this question, refer to the paragraph below, which is a small section of the reading passage on the examination:

> The computer model study for Yucca Mountain included such variables as: *the possibility of earthquakes, predicted water flow through the mountain, increased rainfall due to climate changes, radioactive leakage from the waste containers, and increased temperatures from the buried waste within the containers.* [emphasis added]

The paragraph contains five variables used by the computer model. An incorrect assumption about *any* of these variables would limit the reliability of the model. Some examples of acceptable responses include uncertainties in:

- corrosion rates of the waste containers,
- the possibility of earthquakes, and
- climatic changes that would change the amount of rainfall. [1 point]

69. A leak in one of the containers would cause radioactive waste to enter the ecosystem of the mountain and its surrounding areas. Some acceptable examples of the effects of this leakage might include:

- pollution of fresh water in the area,
- adverse effects on humans, fish, and wildlife,
- the appearance of radioactive substances in the food chain, and
- the contamination of groundwater. [1 point]

70. Some acceptable examples of risks might include:

- some sites might be placed in more populated areas,
- more sites increases the chance of radioactive contamination, and
- more sites increase the risk of terrorism. [1 point]

71. Use Equations 10A and 10B on Reference Table T and Reference Table N to calculate the time needed to reduce the remaining amount to $\frac{1}{32}$ of the original sample:

$$\text{fraction remaining} = \left(\frac{1}{2}\right)^{\frac{t}{T}} = \frac{1}{32} = \left(\frac{1}{2}\right)^5$$

$$\text{number of half-lives} = \frac{t}{T} = \mathbf{5}$$

$$\frac{t}{T} = \frac{t}{30.23 \text{ y}} = 5$$

$$t = (5) \bullet (30.23 \text{ y})$$

$$= \mathbf{151.2 \text{ y}}$$

The answer may be expressed either as 5 half-lives or as any time value between 150 and 152 years. [1 point]

72. According to Reference Table N, the half-lives of cesium-137 and strontium-90 are, respectively, 30.23 and 28.1 years. These half-lives are relatively short. Within a period of 10,000 years, these isotopes would have almost entirely decayed. [1 point]

73. Water may transport the radioactive materials or it may cause the storage containers to corrode.   [1 point]

74. The independent variable is the one that is changed by the student during the experiment. The dependent variable is the one that is measured as a result of the change. The concentration of HCl is the independent variable, and the reaction rate is the dependent variable.   [1 point]

75. Either the temperature or the surface area of the zinc will affect the rate of the reaction and should remain constant during the experiment.

Note that the following terms are equally acceptable for credit: amount of zinc, Zn, concentration of zinc, and (Zn).   [1 point]

76. As the concentration of the HCl increases, the rate will increase due to the increase in the number of collisions among the reactant particles (Zn and HCl).   [1 point]

77. In solution, nitric acid consists almost entirely of hydrogen (H^+ or H_3O^+) and nitrate (NO_3^-) ions. After the acid spill, the hydrogen ion concentration as well as the nitrate ion concentration will have increased in the pond. [1 point]

78. A decrease of 1 pH unit represents a tenfold increase in the hydrogen ion concentration. Therefore, a thousandfold increase will be accompanied by a decrease of 3 pH units. The pH of the pond after the spill will be 5.0. [1 point]

79. Use Reference Table M. Below a pH of 6.0, bromthymol is yellow. [1 point]

Mark (✓) the questions you answered correctly. Count the number of checks and follow the formulas given to determine your score on each topic.

Core Area	☐ Questions Answered Correctly
	41, 68, 74
Section M—Math Skills ☐ Number of checks ÷ 3 × 100 = _____ %	
	2, 3, 4, 5, 11, 47, 51, 52, 67
Section I—Atomic Concepts ☐ Number of checks ÷ 9 × 100 = _____ %	
	1, 6, 36, 38, 53, 54
Section II—Periodic Table ☐ Number of checks ÷ 6 × 100 = _____ %	
	8, 9, 10, 19, 20, 59
Section III—Moles/Stoichiometry ☐ Number of checks ÷ 6 × 100 = _____ %	
	12, 13, 24, 33, 37, 60, 61
Section IV—Chemical Bonding ☐ Number of checks ÷ 7 × 100 = _____ %	
	14, 15, 16, 17, 18, 22, 23, 40, 42, 43, 62, 64, 65
Section V—Physical Behavior of Matter ☐ Number of checks ÷ 13 × 100 = _____ %	
	35, 50, 75, 76
Section VI—Kinetics and Equilibrium ☐ Number of checks ÷ 4 × 100 = _____ %	
	21, 25, 44, 45, 55, 56
Section VII—Organic Chemistry ☐ Number of checks ÷ 6 × 100 = _____ %	
	26, 27, 28, 46, 57, 58, 63
Section VIII—Oxidation-Reduction ☐ Number of checks ÷ 7 × 100 = _____ %	
	29, 30, 31, 48, 66, 77, 78, 79
Section IX—Acids, Bases, and Salts ☐ Number of checks ÷ 8 × 100 = _____ %	
	7, 32, 34, 39, 49, 69, 70, 71, 72, 73
Section X—Nuclear Chemistry ☐ Number of checks ÷ 10 × 100 = _____ %	

Examination August 2003

Chemistry
The Physical Setting

PART A

Answer all questions in this part.

Directions (1-35): For *each* statement or question, write in the answer space the *number* of the word or expression that, of those given, best completes the statement or answers the question. Some questions may require the use of the *Reference Tables for Physical Setting/Chemistry.*

1 Which electron transition represents a gain of energy?

(1) from 2nd to 3rd shell
(2) from 2nd to 1st shell
(3) from 3rd to 2nd shell
(4) from 3rd to 1st shell

1_____

2 Which particles are found in the nucleus of an atom?

(1) electrons, only
(2) neutrons, only
(3) protons and electrons
(4) protons and neutrons

2_____

3 What is the total number of valence electrons in an atom of sulfur in the ground state?

(1) 6   (3) 3
(2) 8   (4) 4   3_____

4 An electron has a charge of

(1) –1 and the same mass as a proton
(2) +1 and the same mass as a proton
(3) –1 and a smaller mass than a proton
(4) +1 and a smaller mass than a proton   4_____

5 The elements in the Periodic Table are arranged in order of increasing

(1) atomic number
(2) atomic radius
(3) mass number
(4) neutron number   5_____

6 What is the correct IUPAC name for the compound NH_4Cl?

(1) nitrogen chloride
(2) nitrogen chlorate
(3) ammonium chloride
(4) ammonium chlorate   6_____

7 Which element is a solid at STP?

(1) H_2 (3) N_2
(2) I_2 (4) O_2 7_____

8 In which compound is the percent by mass of oxygen greatest?

(1) BeO   (3) CaO
(2) MgO   (4) SrO   8_____

9 Based on Reference Table *F*, which of these salts is the best electrolyte?

(1) sodium nitrate
(2) magnesium carbonate
(3) silver chloride
(4) barium sulfate

9_____

10 What is conserved during a chemical reaction?

(1) mass, only
(2) charge, only
(3) both mass and charge
(4) neither mass nor charge

10_____

11 Which type of bond is formed when electrons are transferred from one atom to another?

(1) covalent   (3) hydrogen
(2) ionic   (4) metallic

11_____

12 Which Lewis electron-dot structure is drawn correctly for the atom it represents?

(1) $:\overset{\cdot\cdot}{N}$ (3) $:\overset{\cdot\cdot}{O}:$

(2) $:\overset{\cdot\cdot}{\underset{\cdot\cdot}{F}}:$ (4) $:\overset{\cdot\cdot}{\underset{\cdot\cdot}{Ne}}:$

12_____

13 What occurs when an atom of chlorine forms a chloride ion?

(1) The chlorine atom gains an electron, and its radius becomes smaller.
(2) The chlorine atom gains an electron, and its radius becomes larger.
(3) The chlorine atom loses an electron, and its radius becomes smaller.
(4) The chlorine atom loses an electron, and its radius becomes larger.

13_____

14 Which substance can *not* be decomposed by a chemical change?

(1) Ne   (3) HF

(2) N_2O (4) H_2O 14_____

15 Which of these substances has the strongest intermolecular forces?

(1) H_2O (3) H_2Se

(2) H_2S (4) H_2Te 15_____

16 A real gas behaves more like an ideal gas when the gas molecules are

(1) close and have strong attractive forces between them

(2) close and have weak attractive forces between them

(3) far apart and have strong attractive forces between them

(4) far apart and have weak attractive forces between them   16_____

17 Which phase change is an exothermic process?

(1) $CO_2(s) \rightarrow CO_2(g)$ (3) $Cu(s) \rightarrow Cu(\ell)$

(2) $NH_3(g) \rightarrow NH_3(\ell)$ (4) $Hg(\ell) \rightarrow Hg(g)$ 17_____

18 Which of these contains only one substance?

(1) distilled water   (3) saltwater

(2) sugar water   (4) rainwater   18_____

19 In which group of the Periodic Table do most of the elements exhibit both positive and negative oxidation states?

(1) 17   (3) 12

(2) 2   (4) 7   19_____

20 At the same temperature and pressure, 1.0 liter of $CO(g)$ and 1.0 liter of $CO_2(g)$ have

(1) equal masses and the same number of molecules
(2) different masses and a different number of molecules
(3) equal volumes and the same number of molecules
(4) different volumes and a different number of molecules   20_____

21 Which type of reaction occurs when nonmetal atoms become negative nonmetal ions?

(1) oxidation   (3) substitution
(2) reduction   (4) condensation   21_____

22 Given the reaction:

$$AgCl(s) \underset{}{\overset{H_2O}{\rightleftharpoons}} Ag^+(aq) + Cl^-(aq)$$

Once equilibrium is reached, which statement is accurate?

(1) The concentration of $Ag^+(aq)$ is greater than the concentration of $Cl^-(aq)$.
(2) The $AgCl(s)$ will be completely consumed.
(3) The rates of the forward and reverse reactions are equal.
(4) The entropy of the forward reaction will continue to decrease.   22_____

23 Which structural formula *correctly* represents a hydrocarbon molecule?

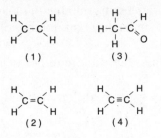

(1) (3) (2) (4)

23 _____

24 Given the structural formulas for two organic compounds:

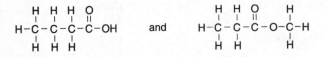

The differences in their physical and chemical properties are primarily due to their different

(1) number of carbon atoms   (3) molecular masses
(2) number of hydrogen atoms   (4) functional groups

24 _____

25 Which structural formula represents a molecule that is *not* an isomer of pentane?

25 _____

26 The bonds in the compound $MgSO_4$ can be described as

(1) ionic, only
(2) covalent, only
(3) both ionic and covalent
(4) neither ionic nor covalent   26_____

27 Given the reaction:

$$Zn(s) + 2\,HCl(aq) \rightarrow ZnCl_2(aq) + H_2(g)$$

Which statement correctly describes what occurs when this reaction takes place in a closed system?

(1) Atoms of $Zn(s)$ lose electrons and are oxidized.
(2) Atoms of $Zn(s)$ gain electrons and are reduced.
(3) There is a net loss of mass.
(4) There is a net gain of mass.   27_____

28 When the pH of a solution changes from a pH of 5 to a pH of 3, the hydronium ion concentration is

(1) 0.01 of the original content
(2) 0.1 of the original content
(3) 10 times the original content
(4) 100 times the original content   28_____

29 A sample of $Ca(OH)_2$ is considered to be an Arrhenius base because it dissolves in water to yield

(1) Ca^{2+} ions as the only positive ions in solution
(2) H_3O^+ ions as the only positive ions in solution
(3) OH^- ions as the only negative ions in solution
(4) H^- ions as the only negative ions in solution   29_____

30 Which reaction occurs when hydrogen ions react with hydroxide ions to form water?

(1) substitution   (3) ionization
(2) saponification   (4) neutralization   30_____

31 Which of these types of nuclear radiation has the greatest penetrating power?

(1) alpha   (3) neutron
(2) beta   (4) gamma   31_____

32 Alpha particles and beta particles differ in

(1) mass, only
(2) charge, only
(3) both mass and charge
(4) neither mass nor charge   32_____

33 Given the nuclear reaction:

$$_{27}^{60}\text{Co} \rightarrow _{-1}^{0}\text{e} + _{28}^{60}\text{Ni}$$

This reaction is an example of

(1) fission
(2) fusion
(3) artificial transmutation
(4) natural transmutation   33_____

34 As two chlorine atoms combine to form a molecule, energy is

(1) absorbed   (3) created
(2) released   (4) destroyed   34_____

Note that question 35 has only three choices.

35 In most aqueous reactions as temperature increases, the effectiveness of collisions between reacting particles

(1) decreases
(2) increases
(3) remains the same   35_____

Part B–1

Answer all questions in this part.

Directions (36-50): For *each* statement or question, write in the answer space the *number* of the word or expression that, of those given, best completes the statement or answers the question. Some questions may require the use of the *Reference Tables for Physical Setting/Chemistry*.

36 What is the total number of neutrons in an atom of an element that has a mass number of 19 and an atomic number of 9?

(1) 9   (3) 19

(2) 10   (4) 28   36_____

37 The element in Period 4 and Group 1 of the Periodic Table would be classified as a

(1) metal   (3) nonmetal

(2) metalloid   (4) noble gas   37_____

38 As the elements in Period 2 of the Periodic Table are considered in succession from left to right, there is a decrease in atomic radius with increasing atomic number. This may best be explained by the fact that the

(1) number of protons increases, and the number of shells of electrons remains the same

(2) number of protons increases, and the number of shells of electrons increases

(3) number of protons decreases, and the number of shells of electrons remains the same

(4) number of protons decreases, and the number of shells of electrons increases   38_____

39 Given the balanced equation:

$$2 C_4H_{10}(g) + 13 O_2(g) \rightarrow 8 CO_2(g) + 10 H_2O(g)$$

What is the total number of moles of $O_2(g)$ that must react completely with 5.00 moles of $C_4H_{10}(g)$?

(1) 10.0   (3) 26.5

(2) 20.0   (4) 32.5   39_____

40 Which particle has the same electron configuration as a potassium ion?

(1) fluoride ion   (3) neon atom

(2) sodium ion   (4) argon atom   40_____

41 Which equation represents a double replacement reaction?

(1) $2 Na + 2 H_2O \rightarrow 2 NaOH + H_2$

(2) $CaCO_3 \rightarrow CaO + CO_2$

(3) $LiOH + HCl \rightarrow LiCl + H_2O$

(4) $CH_4 + 2 O_2 \rightarrow CO_2 + 2 H_2O$ 41_____

42 What is the molecular formula of a compound that has a molecular mass of 54 and the empirical formula C_2H_3?

(1) C_2H_3 (3) C_6H_9

(2) C_4H_6 (4) C_8H_{12} 42_____

43 Given the diagrams X, Y, and Z below:

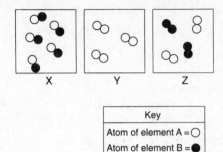

Key
Atom of element A = ○
Atom of element B = ●

Which diagram or diagrams represent a mixture of elements A and B?

(1) X, only   (3) X and Y
(2) Z, only   (4) X and Z   43_____

44 Which is an electron configuration for an atom of chlorine in the excited state?

(1) 2-8-7   (3) 2-8-6-1
(2) 2-8-8   (4) 2-8-7-1   44_____

45 Based on the nature of the reactants in each of the equations below, which reaction at 25°C will occur at the fastest rate?

(1) $C(s) + O_2(g) \rightarrow CO_2(g)$
(2) $NaOH(aq) + HCl(aq) \rightarrow$
    $NaCl(aq) + H_2O(\ell)$
(3) $CH_3OH(\ell) + CH_3COOH(\ell) \rightarrow$
    $CH_3COOCH_3(aq) + H_2O(\ell)$
(4) $CaCO_3(s) \rightarrow CaO(s) + CO_2(g)$ 45_____

46 Given the reaction at equilibrium:

$$A(g) + B(g) \rightleftharpoons AB(g) + heat$$

The concentration of $A(g)$ can be increased by

(1) lowering the temperature
(2) adding a catalyst
(3) increasing the concentration of $AB(g)$
(4) increasing the concentration of $B(g)$ 46_____

47 Which structural formula represents an alcohol?

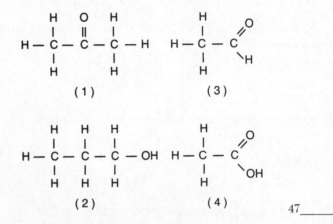

48 A voltaic cell differs from an electrolytic cell in that in a voltaic cell

(1) energy is produced when the reaction occurs
(2) energy is required for the reaction to occur
(3) both oxidation and reduction occur
(4) neither oxidation nor reduction occurs   48_____

49 What is the purpose of the salt bridge in a voltaic cell?

(1) It blocks the flow of electrons.
(2) It blocks the flow of positive and negative ions.
(3) It is a path for the flow of electrons.
(4) It is a path for the flow of positive and negative ions.   49_____

50 According to Reference Table N, which radioactive isotope will retain only one-eighth ($\frac{1}{8}$) its original radioactive atoms after approximately 43 days?

(1) gold-198   (3) phosphorus-32
(2) iodine-131   (4) radon-222   50_____

Part B–2

Answer all questions in this part.

Directions (51-62): Record your answers in the spaces on the answer sheet provided in the back. Some questions may require the use of the *Reference Tables for Physical Setting/Chemistry.*

51 Explain how a catalyst may increase the rate of a chemical reaction. [1]

52 On the set of axes provided *in your answer booklet,* sketch the potential energy diagram for an endothermic chemical reaction that shows the activation energy and the potential energy of the reactants and the potential energy of the products. [2]

53 Given the reaction: $Cl_2 + 2\ HBr \rightarrow Br_2 + 2\ HCl$

Write a correctly balanced reduction half-reaction for this equation. [1]

Base your answers to questions 54 and 55 on the information below.

Given the unbalanced equation:

$$\_\_\ C_6H_{12}O_6 \xrightarrow{\text{enzyme}} \_\_\ C_2H_5OH + \_\_\ CO_2$$

54 Balance the equation provided on the answer sheet, using the lowest whole-number coefficients.   [1]

55 Identify the type of reaction represented.   [1]

_____

Base your answers to questions 56 through 58 on the *Reference Tables for Physical Setting/Chemistry*.

56 Complete the data table provided on the answer sheet for the following Group 18 elements: He, Ne, Ar, Kr, Xe   [1]

57 Using information from your data table in question 56, construct a line graph on the grid provided on the answer sheet, following the directions below.

• Mark an appropriate scale on the axis labeled "First Ionization Energy (kJ/mol)."   [1]
• Plot the data from your data table. Circle each point and connect the points.   [1]

Example: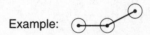

58 Based on your graph in question 57, describe the trend in first ionization energy of Group 18 elements as the atomic number increases.   [1]

Base your answers to questions 59 through 62 on the information below.

Given the heating curve where substance X starts as a solid below its melting point and is heated uniformly:

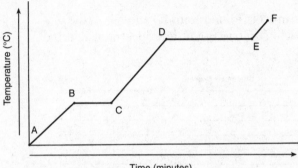

59 Identify the process that takes place during line segment $\overline{DE}$ of the heating curve. [1]

60 Identify a line segment in which the average kinetic energy is increasing. [1]

61 Using (•) to represent particles of substance X, draw at least *five* particles as they would appear in the substance at point F. Use the box provided on the answer sheet. [1]

62 Describe, in terms of *particle behavior* or *energy*, what is happening to substance X during line segment $\overline{BC}$. [1]

Part C

Answer all questions in this part.

Directions (63-78): Record your answers in the spaces on the answer sheet provided in the back. Some questions may require the use of the *Reference Tables for Physical Setting/Chemistry.*

Base your answers to questions 63 and 64 on the diagram below, which shows bright-line spectra of selected elements.

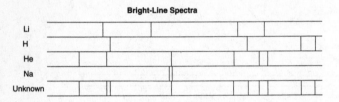

63 Identify the *two* elements in the unknown spectrum. [2]

64 Explain how a bright-line spectrum is produced, in terms of *excited state, energy transitions,* and *ground state.* [2]

65 The table below gives information about two isotopes of element X.

Isotope	Mass	Relative Abundance
X-10	10.01	19.91%
X-11	11.01	80.09%

Calculate the average atomic mass of element X.

- Show a correct numerical setup in the space provided on the answer sheet.   [1]
- Record your answer.   [1]
- Express your answer to the correct number of significant figures.   [1]

66 A student determines the density of zinc to be 7.56
grams per milliliter. If the accepted density is 7.14
grams per milliliter, what is the student's percent
error?

- Show a correct numerical setup in the space provided on the answer sheet.   [1]
- Record your answer.   [1]

Base your answers to questions 67 through 69 on the
information below.

Given the equation for the dissolving of sodium
chloride in water:

$$NaCl(s) \xrightarrow{\text{H}_2\text{O}} Na^+(aq) + Cl^-(aq)$$

67 Describe what happens to entropy during this dissolving process.   [1]

68 Explain, in terms of *particles*, why NaCl(s) does *not* conduct electricity.   [1]

69 When NaCl(s) is added to water in a 250-milliliter beaker, the temperature of the mixture is lower than the original temperature of the water. Describe this observation in terms of *heat flow*.   [1]

Base your answers to questions 70 through 74 on the article below, the *Reference Tables for Physical Setting/Chemistry*, and your knowledge of chemistry.

In the 1920s, paint used to inscribe the numbers on watch dials was composed of a luminescent (glow-in-the-dark) mixture. The powdered-paint base was a mixture of radium salts and zinc sulfide. As the paint was mixed, the powdered base became airborne and drifted throughout the workroom causing the contents of the workroom, including the painters' clothes and bodies, to glow in the dark.

The paint is luminescent because radiation from the radium salts strikes a scintillator. A scintillator is a material that emits visible light in response to ionizing radiation. In watchdial paint, zinc sulfide acts as the scintillator.

Radium present in the radium salts decomposes spontaneously, emitting alpha particles. These particles can cause damage to the body when they enter human tissue. Alpha particles are especially harmful to the blood, liver, lungs, and spleen because they can alter genetic information in the cells. Radium can be deposited in the bones because it substitutes for calcium.

70 Write the notation for the alpha particles emitted by radium in the radium salts.   [1]

71 How can particles emitted from radioactive nuclei damage human tissue? [1]

72 Why does radium substitute for calcium in bones? [1]

73 Explain why zinc sulfide is used in luminescent paint. [1]

74 Based on Reference Table *F*, describe the solubility of zinc sulfide in water. [1]

Base your answers to questions 75 through 78 on the article below and on your knowledge of chemistry.

Fizzies — A Splash from the Past

They're baaack … a splash from the past! Fizzies instant sparkling drink tablets, popular in the 1950s and 1960s, are now back on the market. What sets them apart from other powdered drinks is that they bubble and fizz when placed in water, forming an instant carbonated beverage.

The fizz in Fizzies is caused by bubbles of carbon dioxide (CO_2) gas that are released when the tablet is dropped into water. Careful observation reveals that these bubbles rise to the surface because CO_2 gas is much less dense than water. However, not all of the CO_2 gas rises to the surface; some of it dissolves in the water. The dissolved CO_2 can react with water to form carbonic acid, H_2CO_3.

$$H_2O(\ell) + CO_2(aq) \rightleftharpoons H_2CO_3(aq)$$

The pH of the Fizzies drink registers between 5 and 6, showing that the resulting solution is clearly acidic. Carbonic acid is found in other carbonated beverages as well. One of the ingredients on any soft drink label is carbonated water, which is another name for carbonic acid. However, in the production of soft drinks, the CO_2 is pumped into the solution under high pressure at the bottling plant.

— Brian Rohrig
Excerpted from "Fizzies—A Splash from the Past,"
Chem Matters, February 1998

75 What is the only positive ion in an aqueous solution of carbonic acid?   [1]

76 CO_2 is pumped into the soft drink solution under high pressure. Why is high pressure necessary?   [1]

77 Describe the solubility of CO_2 gas in water.   [1]

78 Explain your response to question 77 in terms of the *molecular polarities* of $CO_2(g)$ and $H_2O(\ell)$.   [1]

Answer Sheet
August 2003

Chemistry
The Physical Setting

Part B–2

Answer all questions in Part B–2 and Part C. Record your answers on the answer sheet.

51 _____

_____

52

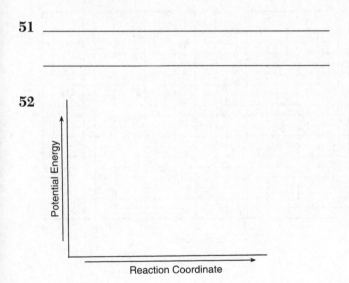

53 _____

54 ___ $C_6H_{12}O_6 \xrightarrow{\text{enzyme}}$ ___ C_2H_5OH + ___ CO_2

55 _____

56

Atomic Number	Element	First Ionization Energy (kJ/mol)
2	He	
10	Ne	
18	Ar	
36	Kr	
54	Xe	

57

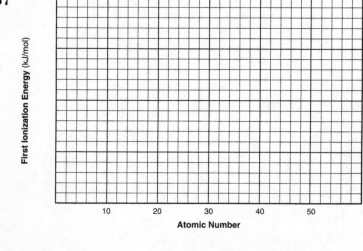

58 _____

_____

59 _____

60 _____

61

62 _____

_____

Part C

63 _____ and _____

64 _____

_____

_____

_____

65

Average atomic mass of element *X:* _____**amu**

66

_____ % error

67 _____

68 _____

_____

69 _____

_____

70 _____

71 _____

_____

72 _____

_____

73 _____

_____

74 _____

_____

75 _____

76 _____

_____

77 _____

78 CO_2: _____

H_2O: _____

Answers
August 2003
Chemistry
The Physical Setting

Answer Key

PART A

1. 1	**7.** 2	**13.** 2	**19.** 1	**25.** 4	**31.** 4
2. 4	**8.** 1	**14.** 1	**20.** 3	**26.** 3	**32.** 3
3. 1	**9.** 1	**15.** 1	**21.** 2	**27.** 1	**33.** 4
4. 3	**10.** 3	**16.** 4	**22.** 3	**28.** 4	**34.** 2
5. 1	**11.** 2	**17.** 2	**23.** 2	**29.** 3	**35.** 2
6. 3	**12.** 4	**18.** 1	**24.** 4	**30.** 4	

PART B–1

36. 2	**40.** 4	**44.** 3	**48.** 1
37. 1	**41.** 3	**45.** 2	**49.** 4
38. 1	**42.** 2	**46.** 3	**50.** 3
39. 4	**43.** 2	**47.** 2	

Answers Explained

PART A

1. **1** When an electron rises to a higher shell, the atom gains energy.

Wrong Choices Explained:
(2), (3), (4) In each of these choices, the electron falls to a lower energy level and the atom loses energy.

2. **4** Protons and neutrons are found in the nucleus; electrons are located in shells outside of the nucleus.

3. **1** Valence electrons are those electrons contained in the outermost shell of the atom. Refer to the Periodic Table of the Elements. Sulfur (S, atomic number 16) has the ground state electron configuration 2–8–6. Therefore, the number of valence electrons is 6.

4. **3** An electron has a negative charge (–1). Its mass is approximately $\frac{1}{2000}$ the mass of a proton.

5. **1** The modern Periodic Table is arranged in order of increasing atomic number because this configuration makes the properties of families of elements more apparent.

6. **3** Refer to Reference Table E. The NH_4^+ ion is named ammonium. The Cl^- ion is named chloride. According to IUPAC rules, the positive ion precedes the negative ion: the proper name is ammonium chloride.

7. **2** At STP, the temperature is 273 K. Refer to Reference Table S. I_2 (iodine, atomic number 53) has a melting point of 387 K. Therefore, I_2 is a solid at STP.

Wrong Choices Explained:
(1), (3), (4) H_2 (hydrogen, atomic number 1), N_2 (nitrogen, atomic number 7), and O_2 (oxygen, atomic number 8) each have a boiling point less than 273 K. Therefore, each of these elements is a gas at STP.

8. **1** Since all of these compounds contain 1 oxygen atom in the formula, the percent oxygen can be found from the relationship:

$$\% \text{ oxygen} = \frac{\text{atomic mass of oxygen (16.0)}}{\text{formula mass of compound}} \times 100\%$$

The compound with the *smallest* formula mass will have the greatest mass percent of oxygen. Since BeO has the smallest formula mass, it has the greatest mass percent of oxygen.

9. **1** An electrolyte is a solution that conducts electricity due to the presence of dissolved ions. Reference Table *F* provides rules for determining the solubility of ionic compounds in water. The compound that is most soluble will be the best electrolyte. Of the choices given, choice (1), NaCl, is the most soluble in water and is therefore the best electrolyte.

Wrong Choices Explained:
(2), (3), (4) According to Reference Table *F*, none of these compounds are appreciably soluble in water.

10. **3** In every chemical reaction, both mass and charge are conserved. For example, in the reaction:

$$H_3O^+ + OH^- \rightarrow 2H_2O$$

the masses on both sides of the reaction (36.0) are equal. In addition, the charge on both sides of the reaction (0) are equal.

11. **2** An ionic bond is formed as the result of a transfer of electrons between two atoms.

Wrong Choices Explained:
(1) A covalent bond is formed as the result of the sharing of one or more pairs of electrons.
(3) A hydrogen bond is a special type of *intermolecular* attraction.
(4) A metallic bond is one formed between the positive ions of metal atoms and allows the valence electrons to move freely throughout the metallic structure.

12. **4** A Lewis electron-dot structure of an atom uses dots to represent the valence electrons of the atom. Of the choices given, only choice (4), Ne, correctly matches the structure with the 8 valence electrons present in the atom.

Wrong Choices Explained:
  (1) Nitrogen has 5 valence electrons; the structure showns only 4 electrons.
  (2) Fluorine has 7 valence electrons; the structure shows 8 electrons.
  (3) Oxygen has 6 valence electrons; the structure shows 7 valence electrons.

13. **2** Chlorine (Cl, atomic number 17) has 7 valence electrons. When an atom of chlorine forms a chloride ion (Cl^-), it gains an electron. Due to the increased repulsion of the valence electrons, the radius of the chloride ion is larger than the radius of the chlorine atom.

14. **1** Elements cannot be decomposed by chemical changes. Of the choices given, only choice (1), Ne, is an element.

15. **1** An atom of O has a higher electronegativity than an atom of S, Se, or Te. As a result, the H_2O molecule is the most polar of the four molecules given in this question. Moreover, an atom of O is smaller than an atom of S, Se, or Te (refer to Reference Table S). As a result, an atom of oxygen in one molecule can approach the hydrogen atoms in another molecule more closely and attract them more strongly than can the other molecules given. The type of intermolecular attractions present among H_2O molecules is known as *hydrogen bonding*.

16. **4** Ideal gas molecules are assumed to be relatively far apart and to have *no* attractive forces between them.

17. **2** When a gas condenses to a liquid, the system's potential energy is reduced. As a result, energy is released into the environment; that is, the phase change is exothermic.

Wrong Choices Explained:
  (1), (3), (4) Sublimation, melting, and boiling are *endothermic* processes.

18. **1** Distilled water is water that has had all other substances removed from it. As a result, distilled water contains only one substance: H_2O.

Wrong Choices Explained:
(2), (3), and (4) Sugar water, saltwater, and even rainwater contain other dissolved substances.

19. **1** Examine the Periodic Table. Of the choices given, only choice (1), Group 17, contains elements that have both positive and negative oxidation states.

20. **3** At the same temperature and pressure, equal volumes of gases contain the same number of molecules (Avogadro's hypothesis). Since the molar masses of CO and CO_2 are 28 grams and 44 grams, respectively, 1.0 liter of CO_2 will have a larger mass than 1.0 liter of CO.

21. **2** A negative ion is formed by *gaining* one or more electrons. The gaining of electrons is known as reduction.

Wrong Choice Explained:
(1) Oxidation involves the *loss* of electrons.

22. **3** Equilibrium is defined as the condition in which the rates of the forward and reverse reactions are equal.

23. **2** In order for a structure to represent a hydrocarbon molecule correctly, the molecule must contain *only* carbon and hydrogen atoms, each carbon atom must be associated with 4 bonds, and each hydrogen atom must be associated with 1 bond. Of the choices given, only choice (2),

meets all three criteria.

24. **4** Refer to Reference Table R. The organic compounds in this question are butanoic acid, an organic acid, and methyl propanoate, an ester. Since the functional groups in these compounds are different, they will have different physical and chemical properties.

25. **4** Isomers of a compound have the same molecular formula but different structural formulas. Refer to Reference Tables P and Q. By using the general formula for alkanes (C_nH_{2n+2}), and setting $n = 5$, we see that the molecular formula for pentane is C_5H_{12}. Now count the number of carbon and hydrogen atoms in each choice. Of the choices given, only choice (4)

does *not* have the molecular formula C_5H_{12}. Therefore, this compound cannot be an isomer of pentane.

26. **3** $MgSO_4$ contains the ions Mg^{2+} and SO_4^{2-}. Therefore it contains ionic bonds. However, within the SO_4^{2-} ion, the oxygen atoms are joined to the sulfur atom by covalent bonds. Therefore, this compound contains both ionic and covalent bonds.

27. **1** The oxidation state of Zn in $Zn(s)$ is 0. In $ZnCl_2(aq)$, the oxidation state of Zn is +2. Atoms increase their oxidation states by *losing* electrons, a process known as oxidation. Therefore, atoms of $Zn(s)$ lose electrons and are oxidized.

Wrong Choices Explained:
(3), (4) Mass is *always* conserved in a chemical reaction.

28. **4** A decrease of 1 pH unit is equivalent to a tenfold (10×) *increase* in the hydronium ion concentration. Therefore, a decrease of 2 pH units reflects a hundredfold (100×) increase in the hydronium ion concentration.

29. **3** An Arrhenius base is defined as a substance that yields OH^- ions as the only negative ions when dissolved in water.

30. **4** Neutralization is the reaction of hydrogen ions and hydroxide ions to form water, as shown below:

$$H^+(aq) + OH^-(aq) \rightarrow H_2O(\ell)$$

31. **4** Gamma radiation has the greatest penetrating power. The order of penetrating power is:

gamma > neutron > beta > alpha

32. **3** Refer to Reference Table *O*. Alpha particles are the nuclei of helium–4 and have positive charges. Beta particles are electrons and have negative charges. In addition, the mass of an alpha particle is much, much greater than the mass of a beta particle. Therefore, alpha and beta particles differ in both charge and mass.

33. **4** The equation represents the natural transmutation of the radioisotope cobalt–60 by beta decay.

34. **2** When a molecule is formed from two separate atoms, the formation of the bond always results in the release of energy.

35. **2** Temperature is a measure of the average kinetic energy of the reacting particles. As the average kinetic energy increases (due to an increase in temperature), the number of *effective* collisions between the reacting particles also increases.

PART B–1

36. **2** The atomic number of an atom is the number of protons contained in the nucleus. The mass number of an atom is the sum of the protons and neutrons contained in the nucleus.

$$\begin{aligned} \text{neutrons} &= \text{mass number} - \text{atomic number} \\ &= 19 - 9 \\ &= \mathbf{10} \end{aligned}$$

37. **1** Refer to the Periodic Table. The element located in Period 4 and Group 1 is potassium (symbol K). Potassium, which lies at the left side of the Periodic Table of the Elements, is one of the most *metallic* elements.

38. **1** As the atomic number increases, the nuclear charge (that is, the number of protons) also increases. Since each additional electron is added to the same (valence) shell, the increased force of attraction between the protons and electrons causes the size of the atom to decrease across the period.

39. **4** The coefficients of the balanced equation indicate the relative number of moles that participate in the reaction: 2 moles of C_4H_{10} react with 13 moles of O_2. Use the factor-label method to determine the answer:

$$5.00 \text{ mol } C_4H_{10} \left(\frac{13 \text{ mol } O_2}{2 \text{ mol } C_4H_{10}} \right) = \textbf{32.5 mol } O_2$$

40. **4** Refer to the Periodic Table of the Elements. An *atom* of potassium (symbol K) has the electron configuration 2-8-8-1. When potassium forms a (positive) *ion*, it loses one electron. Therefore the electron configuration of a potassium ion (K^+) is 2-8-8. Of the choices given only choice (4), argon atom, has an electron configuration of 2-8-8.

Wrong Choices Explained:

(1), (2), (3) A fluoride ion (F^-), a sodium ion (Na^+), and a neon atom (Ne) all have the electron configuration 2-8.

41. **3** In a double replacement reaction, the positive and negative ions "switch" places. Write the equation as though all of the substances were ionic:

$$Li^+[OH]^- + H^+Cl^- \rightarrow L^+Cl^- + H^+[OH]^-$$

Note: H_2O is not really an ionic substance; it is a covalently bonded molecule. It was written this way to illustrate the double replacement.

Wrong Choices Explained:

(1) This equation represents a single replacement reaction.
(2) This equation represents a decomposition reaction.
(4) This equation represents a combustion reaction.

42. **2** Since C_2H_3 is the empirical formula of the compound, the molecular formula must be a multiple of the empirical formula. Additionally, the molecular mass (54) must be a multiple of the empirical formula mass (27). Since the molecular mass is twice the empirical formula mass, the molecular formula must be twice that of the empirical formula, or C_4H_6.

43. **2** In a mixture, some of the substance will not be chemically combined. Diagram Z illustrates a mixture of elements A and B, present as A_2 and B_2 molecules.

Wrong Choices Explained:
(1) Diagram X represents the *compound AB*.
(3) Diagram Y represents the element A, present as A_2 molecules.

44. **3** Refer to the Periodic Table of the Elements. The ground state electron configuration of an atom of chlorine (symbol Cl) is 2-8-7. In an excited state, one or more of the electrons will be promoted to higher shells, but the total number of electrons *must* remain at 17. Of the choices given, only choice (3), 2-8-6-1, meets both of these criteria.

45. **2** Systems containing dissolved *ions* will react very rapidly since no bonds need to be broken. In this reaction, the ions $Na^+(aq)$, $OH^-(aq)$, $H^+(aq)$, and $Cl^-(aq)$ undergo a double replacement at a very rapid rate.

46. **3** According to LeChâtelier's principle, an increase in concentration of the product (AB) will shift the reaction so that some of the product is removed. In other words, the reaction will shift to the left. As a result, the concentration of A (and of B) will increase.

Wrong Choices Explained:
(1) Lowering the temperature will favor the *exothermic* reaction. The reaction will shift toward the right, leading to a decrease in the concentration of A.
(2) Adding a catalyst will not shift the reaction in either direction.
(4) Increasing the concentration of B will shift the reaction toward the right, leading to a decrease in the concentration of A.

47. **2** Refer to Reference Table R. An alcohol contains the functional group –NOH.

Wrong Choices Explained:
(1) This compound is a ketone, known as propanone.
(3) This compound is an aldehyde, known as ethanal.
(4) This compound is an organic acid, known as ethanoic acid.

48. **1** In a voltaic cell, a redox reaction is used to produce electrical energy. In an electrolytic cell, a source of electrical energy is used to drive a redox reaction.

49. **4** In a voltaic cell, electrons are transferred through the external circuit (that is, the wire connecting the half-cells). In order to *complete* the circuit, a flow of ions must be maintained. In some types of voltaic cells, this is accomplished by means of a salt bridge.

50. **3** Use Reference Table *N* and Equations *10A* and *10B* on Reference Table *T*:

$$\left(\frac{1}{2}\right)^{\frac{t}{T}} = \frac{1}{8}$$

$$\frac{t}{T} = 3 \text{ half-lives}$$

Therefore, since 43 days equals 3 half-lives, the half-life of the radioisotope must equal $\frac{43}{3}$ days = 14.3 days. An inspection of Reference Table *N* shows that phosphorous-32 (^{32}P) has a half-life of 14.3 days.

PART B–2

[Point values are indicated in brackets.]

51. Catalysts increase the rate of a reaction by lowering the activation energy of the reaction. Other acceptable responses include (1) providing an alternate reaction pathway and (2) forming a different activated complex with a lower activation energy.   [1 point]

52.

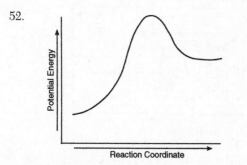

Note: One point is allowed for showing that the peak (the activated complex) is the *highest* point in the curve, and one point is allowed for showing that the end of the curve is higher than the beginning of the curve.   [2 points]

53. Reduction is the gain of electrons. The half-reaction must be a balanced equation that shows atoms, ions, and electrons. Acceptable responses include:

$$Cl_2 + 2e^- \rightarrow 2Cl^-$$
$$Cl_2 \rightarrow 2Cl^- - 2e^-$$
$$Cl + e^- \rightarrow Cl^-$$

[1 point]

54. The correctly balanced equation is $C_6H_{12}O_6 \rightarrow \mathbf{2}C_2H_5OH + \mathbf{2}CO_2$. Including a coefficient of **1** in front of the $C_6H_{12}O_6$ is acceptable.   [1 point]

55. Acceptable responses include fermentation, decomposition, or redox. [1 point]

56.

Atomic Number	Element	First Ionization Energy (kJ/mol)
2	He	2372
10	Ne	2081
18	Ar	1521
36	Kr	1351
54	Xe	1170

Note: All values must be correct in order to receive credit.   [1 point]

57.

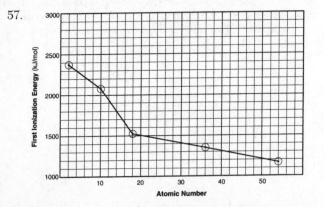

Note: One point is awarded for marking an appropriate scale on the vertical axis, and one point is allowed for correctly plotting and connecting the data points.   [2 points]

58. As the atomic number increases within Group 18, the first ionization energy *decreases*. [1 point]

59–62. Refer to the heating curve shown below:

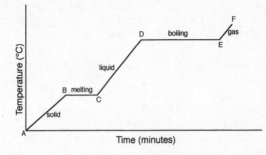

Time (minutes)

59. Between points D and E, liquid is changing to gas, that is, boiling is occurring. Other acceptable responses include vaporization and liquid-vapor equilibrium. [1 point]

60. As the temperature rises, the average kinetic energy of the molecules increases. Select any region in which the temperature is rising: AB, CD, or EF. [1 point]

61. The substance is a gas at point F. An acceptable particle diagram is shown below:

[1 point]

62. Melting is occurring over region BC. All of the following events are occurring and any one is acceptable for credit:

- The average kinetic energy of the particles remains the same.
- The potential energy of the particles is increasing.
- The particles are becoming more disordered.
- The distance between the particles is increasing.
- The intermolecular forces of attraction are decreasing. [1 point]

PART C

[Point values are indicated in brackets.]

63. Compare the positions of the lines in the unknown sample with the positions of the lines in the elements given above it. The unknown sample is a mixture of hydrogen (H) and helium (He).

Note: One point is awarded for identifying each of the elements present. [2 points]

64. In order to produce a bright-line spectrum, energy must be absorbed by atoms, raising the electrons from the ground state to one or more excited states. As the electrons return to lower energy levels (shells), the energy that is released forms the bright-line spectrum.

Note: One point is awarded for explaining the change from the excited state to the ground state, one point is awarded for explaining the release of energy. [2 points]

65. To calculate the average atomic mass of an element, multiply the mass of each naturally occurring isotope by the *decimal equivalent* of the relative abundance and then add the results:

$$(0.1991) \bullet (10.01) + (0.8009) \bullet (11.01) = \mathbf{10.81}$$

Note: One point is awarded for the correct setup, one point is awarded for the correct answer, and one point is awarded for expressing the answer to the correct number of significant figures.   [3 points]

66. The percent error of a measurement is calculated from the following relationship:

$$\% \text{ error} = \frac{(\text{experimental value}) - (\text{accepted value})}{(\text{accepted value})} \times 100\%$$

$$\% \text{ error} = \frac{7.56 \text{ g/mL} - 7.14 \text{ g/mL}}{7.14 \text{ g/mL}} \times 100\% = \mathbf{5.9\%}$$

Note: One point is awarded for the correct numerical setup, and one point is awarded for the correct answer (5.9%, or 5.88%, or 6%).   [2 points]

67. As the solid NaCl dissolves in water, the dissolved ions are distributed more randomly than they were in the crystal. That is, the entropy of the system increases. [1 point]

68. Solid NaCl does not conduct electricity because the ions in the crystal are not mobile. [1 point]

69. The dissolving of NaCl in water is an endothermic process in which heat is absorbed by the system (the dissolved ions) from its surroundings (the water). This inward flow of energy leads to a lowering in the temperature. [1 point]

70. Refer to Reference Table O. The symbol for an alpha particle is ^4_2He or $^4_2\alpha$. [1 point]

71. Particles emitted from radioactive nuclei can damage human tissue by altering the genetic information in the cells. [1 point]

72. Refer to the Periodic Table of the Elements. Radium (symbol Ra) and calcium (symbol Ca) are both located within Group 2 of the Periodic Table of the Elements. Therefore, both elements will possess similar chemical properties. [1 point]

73. The zinc sulfide is a scintillator, that is, a substance that emits visible light when exposed to ionizing radiation. [1 point]

74. Refer to Reference Table F. Since sulfides are generally insoluble in water and Zn^{2+} is not a Group 1 ion, it is expected that zinc sulfide will *not* be soluble in water. [1 point]

75. Arrhenius acids, such as carbonic acid, release hydronium ions (H^+ or H_3O^+) as the only positive ion in aqueous solution. [1 point]

76. The solubility of a gas, such as CO_2, increases in water when the pressure on the gas is increased. [1 point]

77. CO_2 has a low solubility in water. [1 point]

78. Due to its symmetry, CO_2 is a nonpolar molecule. In contrast, H_2O is polar because the molecule is not symmetrical. In general, molecules need to have similar polarities in order to form a solution. [1 point]

Mark (✓) the questions you answered correctly. Count the number of checks and follow the formulas given to determine your score on each topic.

Core Area	☐ Questions Answered Correctly

57, 58, 65,66

Section M—Math Skills
☐ Number of checks ÷ 4 × 100 = _____ %

1, 2, 3, 4, 36, 44, 63, 64, 65

Section I—Atomic Concepts
☐ Number of checks ÷ 9 × 100 = _____ %

5, 7, 19, 37, 38, 56, 72

Section II—Periodic Table
☐ Number of checks ÷ 7 × 100 = _____ %

6, 8, 10, 39, 41, 42, 54, 55

Section III—Moles/Stoichiometry
☐ Number of checks ÷ 8 × 100 = _____ %

11, 12, 13, 26, 34, 40, 73

Section IV—Chemical Bonding
☐ Number of checks ÷ 7 × 100 = _____ %

14, 15, 16, 17, 18, 20, 43, 59, 60, 61, 62, 68, 69, 74, 76, 77, 78

Section V—Physical Behavior of Matter
☐ Number of checks ÷ 17 × 100 = _____ %

22, 35, 45, 46, 51, 52, 67

Section VI—Kinetics and Equilibrium
☐ Number of checks ÷ 7 × 100 = _____ %

23, 24, 25, 47

Section VII—Organic Chemistry
☐ Number of checks ÷ 4 × 100 = _____ %

21, 27, 48, 49, 53

Section VIII—Oxidation-Reduction
☐ Number of checks ÷ 5 × 100 = _____ %

9, 28, 29, 30, 75

Section IX—Acids, Bases, and Salts
☐ Number of checks ÷ 5 × 100 = _____ %

31, 32, 33, 50, 70, 71

Section X—Nuclear Chemistry
☐ Number of checks ÷ 6 × 100 = _____ %

Examination
June 2004
Chemistry
The Physical Setting

PART A

Answer all questions in this part.

Directions (1–33): For *each* statement or question, write in the answer space the *number* of the word or expression that, of those given, best completes the statement or answers the question. Some questions may require the use of the *Reference Tables for Physical Setting/Chemistry.*

1 The modern model of the atom is based on the work of

   (1) one scientist over a short period of time
   (2) one scientist over a long period of time
   (3) many scientists over a short period of time
   (4) many scientists over a long period of time   1_____

2 Which statement is true about the charges assigned to an electron and a proton?

   (1) Both an electron and a proton are positive.
   (2) An electron is positive and a proton is negative.
   (3) An electron is negative and a proton is positive.
   (4) Both an electron and a proton are negative.   2_____

3 In the wave-mechanical model, an orbital is a region of space in an atom where there is

   (1) a high probability of finding an electron
   (2) a high probability of finding a neutron
   (3) a circular path in which electrons are found
   (4) a circular path in which neutrons are found   3_____

4 What is the charge of the nucleus in an atom of oxygen-17?

   (1) 0   (3) +8
   (2) −2   (4) +17   4_____

5 Which pair of symbols represent a metalloid and a noble gas?

   (1) Si and Bi   (3) Ge and Te
   (2) As and Ar   (4) Ne and Xe   5_____

6 Which statement describes a chemical property of iron?

   (1) Iron can be flattened into sheets.
   (2) Iron conducts electricity and heat.
   (3) Iron combines with oxygen to form rust.
   (4) Iron can be drawn into a wire.   6_____

7 Given the reaction:

$$N_2(g) + 3\ H_2(g) \rightleftharpoons 2\ NH_3(g)$$

   What is the mole-to-mole ratio between nitrogen gas and hydrogen gas?

   (1) 1:2   (3) 2:2
   (2) 1:3   (4) 2:3   7_____

8 What is the percent by mass of oxygen in propanal, CH_3CH_2CHO?

(1) 10.0%   (3) 38.1%

(2) 27.6%   (4) 62.1%   8_____

9 Covalent bonds are formed when electrons are

(1) transferred from one atom to another
(2) captured by the nucleus
(3) mobile within a metal
(4) shared between two atoms   9_____

10 Which type of molecule is CF_4?

(1) polar, with a symmetrical distribution of charge
(2) polar, with an asymmetrical distribution of charge
(3) nonpolar, with a symmetrical distribution of charge
(4) nonpolar, with an asymmetrical distribution of charge   10_____

11 Which change occurs when a barium atom loses two electrons?

(1) It becomes a negative ion and its radius decreases.
(2) It becomes a negative ion and its radius increases.
(3) It becomes a positive ion and its radius decreases.
(4) It becomes a positive ion and its radius increases.   11_____

12 Conductivity in a metal results from the metal atoms having

(1) high electronegativity
(2) high ionization energy
(3) highly mobile protons in the nucleus
(4) highly mobile electrons in the valence shell   12_____

13 Which of these elements has the *least* attraction for electrons in a chemical bond?

(1) oxygen   (3) nitrogen
(2) fluorine   (4) chlorine   13_____

14 Recovering the salt from a mixture of salt and water could best be accomplished by

(1) evaporation
(2) filtration
(3) paper chromatography
(4) density determination   14_____

15 The average kinetic energy of water molecules is greatest in which of these samples?

(1) 10 g of water at 35°C
(2) 10 g of water at 55°C
(3) 100 g of water at 25°C
(4) 100 g of water at 45°C   15_____

16 Helium is most likely to behave as an ideal gas when it is under

(1) high pressure and high temperature
(2) high pressure and low temperature
(3) low pressure and high temperature
(4) low pressure and low temperature   16_____

17 At STP, the element oxygen can exist as either O_2 or O_3 gas molecules. These two forms of the element have

(1) the same chemical and physical properties
(2) the same chemical properties and different physical properties
(3) different chemical properties and the same physical properties
(4) different chemical and physical properties   17_____

18 Which sample contains particles in a rigid, fixed, geometric pattern?

(1) $CO_2(aq)$ (3) $H_2O(\ell)$

(2) $HCl(g)$ (4) $KCl(s)$ 18_____

19 Given the reaction at 25°C:

$$Zn(s) + 2\ HCl(aq) \rightarrow ZnCl_2(aq) + H_2(g)$$

The rate of this reaction can be increased by using 5.0 grams of powdered zinc instead of a 5.0-gram strip of zinc because the powdered zinc has

(1) lower kinetic energy
(2) lower concentration
(3) more surface area
(4) more zinc atoms   19_____

20 Which statement about a system at equilibrium is true?

(1) The forward reaction rate is less than the reverse reaction rate.
(2) The forward reaction rate is greater than the reverse reaction rate.
(3) The forward reaction rate is equal to the reverse reaction rate.
(4) The forward reaction rate stops and the reverse reaction rate continues.   20_____

21 A catalyst increases the rate of a chemical reaction by

(1) lowering the activation energy of the reaction
(2) lowering the potential energy of the products
(3) raising the temperature of the reactants
(4) raising the concentration of the reactants   21_____

22 Which element must be present in an organic compound?

(1) hydrogen   (3) carbon
(2) oxygen   (4) nitrogen   22_____

23 Which compound is a saturated hydrocarbon?

(1) hexane   (3) hexanol
(2) hexene   (4) hexanal   23_____

24 Given the reaction:

$$CH_3\overset{\overset{O}{\|}}{C}-OH + HOC_2H_5 \rightleftharpoons CH_3\overset{\overset{O}{\|}}{C}-O-C_2H_5 + H_2O$$

This reaction is an example of

(1) fermentation   (3) hydrogenation
(2) saponification   (4) esterification   24_____

25 Which of these compounds has chemical properties most similar to the chemical properties of ethanoic acid?

(1) C_3H_7COOH (3) $C_2H_5COOC_2H_5$
(2) C_2H_5OH (4) $C_2H_5OC_2H_5$ 25_____

26 Given the reaction that occurs in an electrochemical cell:

$$Zn(s) + CuSO_4(aq) \rightarrow ZnSO_4(aq) + Cu(s)$$

During this reaction, the oxidation number of Zn changes from

(1) 0 to +2   (3) +2 to 0
(2) 0 to −2   (4) −2 to 0   26_____

27 A voltaic cell spontaneously converts

   (1) electrical energy to chemical energy
   (2) chemical energy to electrical energy
   (3) electrical energy to nuclear energy
   (4) nuclear energy to electrical energy   27_____

28 Which pair of formulas represents two compounds that are electrolytes?

   (1) HCl and CH_3OH
   (2) HCl and NaOH
   (3) C_5H_{12} and CH_3OH
   (4) C_5H_{12} and NaOH   28_____

29 Hydrogen chloride, HCl, is classified as an Arrhenius acid because it produces

   (1) H^+ ions in aqueous solution
   (2) Cl^- ions in aqueous solution
   (3) OH^- ions in aqueous solution
   (4) NH_4^+ ions in aqueous solution   29_____

30 Which compound could serve as a reactant in a neutralization reaction?

   (1) NaCl   (3) CH_3OH
   (2) KOH   (4) CH_3CHO 30_____

31 Which of these particles has the greatest mass?

   (1) alpha   (3) neutron
   (2) beta   (4) positron   31_____

32 In a nuclear fusion reaction, the mass of the products is

(1) less than the mass of the reactants because some of the mass has been converted to energy

(2) less than the mass of the reactants because some of the energy has been converted to mass

(3) more than the mass of the reactants because some of the mass has been converted to energy

(4) more than the mass of the reactants because some of the energy has been converted to mass   32_____

33 Which of these types of radiation has the greatest penetrating power?

(1) alpha   (3) gamma

(2) beta   (4) positron   33_____

PART B–1

Answer all questions in this part.

Directions (34–50): For *each* statement or question, write in the answer space the *number* of the word or expression that, of those given, best completes the statement or answers the question. Some questions may require the use of the *Reference Tables for Physical Setting/Chemistry.*

34 How many electrons are contained in an Au^{3+} ion?

(1) 76   (3) 82

(2) 79   (4) 197   34_____

35 Which electron configuration represents the electrons of an atom in an excited state?

(1) 2–4   (3) 2–7–2

(2) 2–6   (4) 2–8–2   35_____

36 In comparison to an atom of $^{19}_{9}F$ in the ground state, an atom of $^{12}_{6}C$ in the ground state has

(1) three fewer neutrons

(2) three fewer valence electrons

(3) three more neutrons

(4) three more valence electrons   36_____

37 Element X is a solid that is brittle, lacks luster, and has six valence electrons. In which group on the Periodic Table would element X be found?

(1) 1   (3) 15

(2) 2   (4) 16   37_____

38 What is the empirical formula for the compound $C_6H_{12}O_6$?

(1) CH_2O (3) $C_3H_6O_3$

(2) $C_2H_4O_2$ (4) $C_6H_{12}O_6$ 38_____

39 The bonds between hydrogen and oxygen in a water molecule are classified as

(1) polar covalent
(2) nonpolar covalent
(3) ionic
(4) metallic 39_____

40 The graph below represents the uniform heating of a substance, starting with the substance as a solid below its melting point.

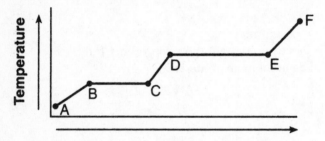

Time

Which line segment represents an increase in potential energy and no change in average kinetic energy?

(1) $\overline{AB}$ (3) $\overline{CD}$
(2) $\overline{BC}$ (4) $\overline{EF}$ 40_____

41 Using your knowledge of chemistry and the information in Reference Table *H*, which statement concerning propanone and water at 50°C is true?

(1) Propanone has a higher vapor pressure and stronger intermolecular forces than water.
(2) Propanone has a higher vapor pressure and weaker intermolecular forces than water.
(3) Propanone has a lower vapor pressure and stronger intermolecular forces than water.
(4) Propanone has a lower vapor pressure and weaker intermolecular forces than water. 41_____

42 A solution that is at equilibrium must be

   (1) concentrated   (3) saturated

   (2) dilute   (4) unsaturated   42_____

43 Given the reaction:

$$N_2(g) + O_2(g) + 182.6 \text{ kJ} \rightleftharpoons 2 \text{ NO}(g)$$

Which change would cause an immediate increase in the rate of the forward reaction?

   (1) increasing the concentration of $NO(g)$

   (2) increasing the concentration of $N_2(g)$

   (3) decreasing the reaction temperature

   (4) decreasing the reaction pressure   43_____

44 Which 10-milliliter sample of water has the greatest degree of disorder?

   (1) $H_2O(g)$ at 120°C

   (2) $H_2O(\ell)$ at 80°C

   (3) $H_2O(\ell)$ at 20°C

   (4) $H_2O(s)$ at 0°C   44_____

45 Which pH indicates a basic solution?

   (1) 1   (3) 7

   (2) 5   (4) 12   45_____

46 Which structural formula represents 2-pentyne?

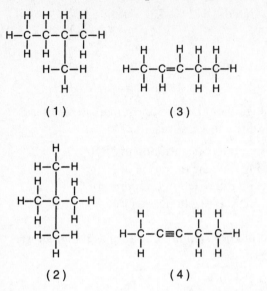

(1)

(2)

(3)

(4)

46_____

47 Which structural formula represents an ether?

(1)

(2)

(3)

(4)

47_____

48 Given the reaction for the corrosion of aluminum:

$$4 Al + 3 O_2 \rightarrow 2 Al_2O_3$$

Which half-reaction correctly represents the oxidation that occurs?

(1) $Al + 3e^- \rightarrow Al^{3+}$
(2) $Al \rightarrow Al^{3+} + 3e^-$
(3) $O_2 + 4e^- \rightarrow 2 O^{2-}$
(4) $O_2 \rightarrow 2 O^{2-} + 4e^-$

48_____

49 Based on Reference Table N, what fraction of a sample of gold-198 remains radioactive after 2.69 days?

(1) $\frac{1}{4}$ (3) $\frac{3}{4}$

(2) $\frac{1}{2}$ (4) $\frac{7}{8}$

49_____

Note that question 50 has only three choices.

50 As the elements of Group 1 on the Periodic Table are considered in order of increasing atomic radius, the ionization energy of each successive element generally

(1) decreases
(2) increases
(3) remains the same

50_____

PART B–2

Answer all questions in the part.

Directions (51–64): Record your answers on the answer sheet provided in the back. Some questions may require the use of the *Reference Tables for Physical Setting/Chemistry*.

Base your answers to questions 51 through 53 on the balanced chemical equation below.

$$2 H_2O \rightarrow 2 H_2 + O_2$$

51 What type of reaction does this equation represent? [1]

52 How does the balanced chemical equation show the Law of Conservation of Mass? [1]

53 What is the total number of moles of O_2 produced when 8 moles of H_2O is completely consumed? [1]

Base your answers to questions 54 and 55 on the unbalanced redox reaction below.

$$Cu(s) + AgNO_3(aq) \rightarrow Cu(NO_3)_2(aq) + Ag(s)$$

54 Write the reduction half-reaction. [1]

55 Balance the redox equation *on the answer sheet*, using the smallest whole-number coefficients. [1]

Base your answers to questions 56 through 58 on the information below.

A student titrates 60.0 mL of HNO_3(aq) with 0.30 M NaOH(aq). Phenolphthalein is used as the indicator. After adding 42.2 mL of NaOH(aq), a color change remains for 25 seconds, and the student stops the titration.

56 What color change does phenolphthalein undergo during this titration? [1]

57 In the space provided *on the answer sheet*, show a correct numerical setup for calculating the molarity of the HNO_3(aq). [1]

58 According to the data, how many significant figures should be present in the calculated molarity of the HNO_3(aq)? [1]

Base your answers to questions 59 through 61 on the data table below, which shows three isotopes of neon.

Isotope	Atomic Mass (atomic mass units)	Percent Natural Abundance
^{20}Ne	19.99	90.9%
^{21}Ne	20.99	0.3%
^{22}Ne	21.99	8.8%

59 In terms of *atomic particles*, state one difference between these three isotopes of neon. [1]

60 Based on the atomic masses and the natural abundances shown in the data table, in the space provided *on the answer sheet*, show a correct numerical setup for calculating the average atomic mass of neon.   [1]

61 Based on natural abundances, the average atomic mass of neon is closest to which whole number? [1]

62 Based on the Periodic Table, explain why Na and K have similar chemical properties.   [1]

63 In the space to the right of the reactants and arrow provided *on the answer sheet*, draw the structural formula for the product of the reaction shown.   [1]

64 Given the nuclear equation:

$$^{58}_{29}\text{Cu} \rightarrow\ ^{58}_{28}\text{Ni} + X$$

What nuclear particle is represented by X?   [1]

PART C

Answer all questions in this part.

Directions (65–85): Record your answers on the answer sheet provided in the back. Some questions may require the use of the *Reference Tables for Physical Setting/Chemistry.*

Base your answers to questions 65 through 67 on the information and equation below.

Antacids can be used to neutralize excess stomach acid. Brand *A* antacid contains the acid-neutralizing agent magnesium hydroxide, $Mg(OH)_2$. It reacts with $HCl(aq)$ in the stomach, according to the following balanced equation:

$$2\ HCl(aq) + Mg(OH)_2(s) \rightarrow MgCl_2(aq) + 2\ H_2O(\ell)$$

65 In the space provided *on the answer sheet,* show a correct numerical setup for calculating the number of moles of $Mg(OH)_2$ (gram-formula mass = 58.3 grams/mole) in an 8.40-gram sample. [1]

66 If a person produces 0.050 mole of excess HCl in the stomach, how many moles of $Mg(OH)_2$ are needed to neutralize this excess hydrochloric acid? [1]

67 Brand *B* antacid contains the acid-neutralizing agent sodium hydrogen carbonate. Write the chemical formula for sodium hydrogen carbonate. [1]

Base your answers to questions 68 through 70 on the information below.

Naphthalene, a nonpolar substance that sublimes at room temperature, can be used to protect wool clothing from being eaten by moths.

68 Explain, in terms of *intermolecular forces*, why naphthalene sublimes.   [1]

69 Explain why naphthalene is *not* expected to dissolve in water.   [1]

70 The empirical formula for naphthalene is C_5H_4 and the molecular mass of naphthalene is 128 grams/mole. What is the molecular formula for naphthalene?   [1]

Base your answers to questions 71 through 74 on the data table below, which shows the solubility of a solid solute.

The Solubility of the Solute at Various Temperatures

Temperature (°C)	Solute per 100 g of H_2O(g)
0	18
20	20
40	24
60	29
80	36
100	49

71 On the grid provided *on the answer sheet*, mark an appropriate scale on the axis labeled "Solute per 100 g of H$_2$O(g)." An appropriate scale is one that allows a trend to be seen. [1]

72 On the same grid, plot the data from the data table. Circle and connect the points. [1]

Example:

73 Based on the data table, if 15 grams of solute is dissolved in 100 grams of water at 40°C, how many *more* grams of solute can be dissolved in this solution to make it saturated at 40°C? [1]

74 According to Reference Table *G*, how many grams of KClO$_3$ must be dissolved in 100 grams of H$_2$O at 10°C to produce a saturated solution? [1]

Base your answers to questions 75 through 78 on the information below.

A weather balloon has a volume of 52.5 liters at a temperature of 295 K. The balloon is released and rises to an altitude where the temperature is 252 K.

75 How does this temperature change affect the gas particle motion? [1]

76 The original pressure at 295 K was 100.8 kPa and the pressure at the higher altitude at 252 K is 45.6 kPa. Assume the balloon does not burst. In the space provided *on the answer sheet*, show a correct numerical setup for calculating the volume of the balloon at the higher altitude.   [1]

77 What Celsius temperature is equal to 252 K?   [1]

78 What pressure, in atmospheres (atm), is equal to 45.6 kPa?   [1]

Base your answers to questions 79 and 80 on the information and equation below.

Human blood contains dissolved carbonic acid, H_2CO_3, in equilibrium with carbon dioxide and water. The equilibrium system is shown below.

$$H_2CO_3(aq) \rightleftharpoons CO_2(aq) + H_2O(\ell)$$

79 Explain, using LeChatelier's principle, why decreasing the concentration of CO_2 decreases the concentration of H_2CO_3.   [1]

80 What is the oxidation number of carbon in $H_2CO_3(aq)$?   [1]

Base your answers to questions 81 through 84 on the information below.

A safe level of fluoride ions is added to many public drinking water supplies. Fluoride ions have been found to help prevent tooth decay. Another common source of fluoride ions is toothpaste. One of the fluoride compounds used in toothpaste is tin(II) fluoride.

A town located downstream from a chemical plant was concerned about fluoride ions from the plant leaking into its drinking water. According to the Environmental Protection Agency, the fluoride ion concentration in drinking water cannot exceed 4 ppm. The town hired a chemist to analyze its water. The chemist determined that a 175-gram sample of the town's water contains 0.000 250 gram of fluoride ions.

81 In the box provided *on the answer sheet*, draw a Lewis electron-dot diagram for a fluoride ion. [1]

82 What is the chemical formula for tin(II) fluoride? [1]

83 How many parts per million of fluoride ions are present in the analyzed sample? [1]

84 Is the town's drinking water safe to drink? Support your decision using information in the passage and your calculated fluoride level in question 83. [1]

---

85 A plan is being developed for an experiment to test the effect of concentrated strong acids on a metal surface protected by various coatings. Some safety precautions would be the wearing of chemical safety goggles, an apron, and gloves. State one additional safety precaution that should be included in the plan. [1]

284

Answer Sheet
June 2004

Chemistry
The Physical Setting

PART B–2

51 _____

52 _____

    _____

53 _____ mol

54 _____

55 _____ Cu(s) + _____ AgNO$_3$(aq) → _____ Cu(NO$_3$)$_2$(aq) +
    _____ Ag(s)

56 _____ to _____

57

58 _____

59 _____

_____

60

61 _____

62 _____

_____

_____

63

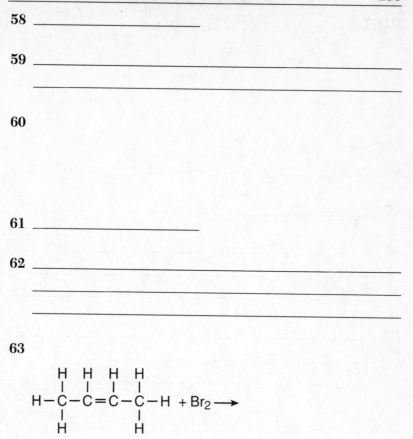

64 _____

PART C

65

66 _____ mol

67 _____

68 _____
_____
_____

69 _____
_____
_____

70 _____

71 and 72

Solubility Curve

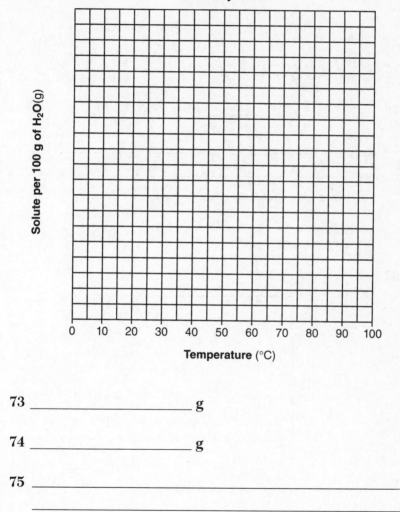

Solute per 100 g of H₂O(g)

Temperature (°C)

73 _____ **g**

74 _____ **g**

75 _____

_____

76

77 _____ °C

78 _____ atm

79 _____

_____

_____

80 _____

81

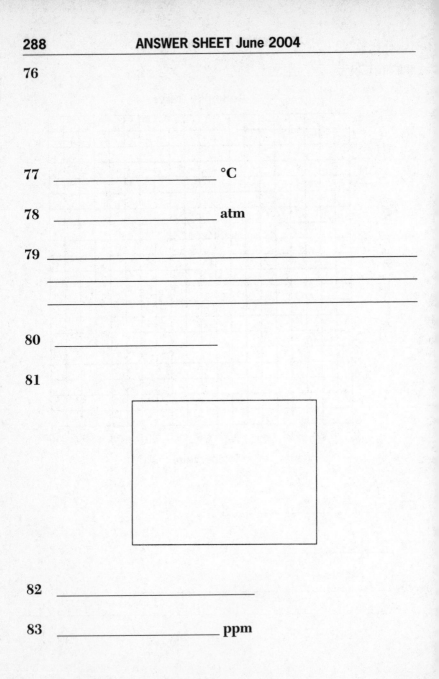

82 _____

83 _____ ppm

84 _____

_____

_____

85 _____

_____

Answers
June 2004

Chemistry
The Physical Setting

Answer Key

PART A

1. 4	8. 2	15. 2	22. 3	29. 1
2. 3	9. 4	16. 3	23. 1	30. 2
3. 1	10. 3	17. 4	24. 4	31. 1
4. 3	11. 3	18. 4	25. 1	32. 1
5. 2	12. 4	19. 3	26. 1	33. 3
6. 3	13. 3	20. 3	27. 2	
7. 2	14. 1	21. 1	28. 2	

PART B–1

34. 1	38. 1	42. 3	45. 4	48. 2
35. 3	39. 1	43. 2	46. 4	49. 2
36. 2	40. 2	44. 1	47. 4	50. 1
37. 4	41. 2			

Answers Explained

PART A

1. **4** The model of the atom began with the work of Dalton in the early nineteenth century and continued into the twentieth century with the work of Thomson, Rutherford, Bohr, and many others. Even today, in the twenty-first century, work on the atomic model continues.

2. **3** The electron is the basic particle that carries a negative charge. The proton is the basic particle that carries a positive charge.

3. **1** An orbital is defined as the region of space in which an electron in an atom is most probably located.

4. **3** The atomic number of an atom is equal to the number of protons in the nucleus of that atom. Refer to the Periodic Table of the Elements. Since oxygen has an atomic number of 8, its nuclear charge is +8.

5. **2** Refer to the Periodic Table of the Elements. Metalloids are found along the metal-nonmetal line in Groups 13, 14, 15, and 16. Noble gases are found in Group 18. Of the choices given, only choice (2), As and Ar, meet these criteria.

Wrong Choices Explained:
(1) Si is a metalloid, but Bi is not a noble gas.
(3) Ge and Te are both metalloids.
(4) Ne and Xe are both noble gases.

6. **3** When a chemical property of a substance is measured or observed, the substance undergoes a *chemical change*. In other words, a chemical reaction occurs. Of the four choices given, only choice (3), "Iron combines with oxygen to form rust," represents a chemical reaction.

Wrong Choices Explained:
(1), (2), (4) Each of these are *physical properties*. When a physical property of a substance is measured or observed, only a physical change occurs. In each of these cases, the iron remains a metallic element.

7. **2** The coefficients of a balanced chemical reaction give the mole ratios of the reactants and products. Since the coefficients of N_2 and H_2 are 1 and 3, respectively, the mole-to-mole ratio must be 1:3.

8. **2** First complete the following table for CH_3CH_2CHO:

Element	Atomic Mass	Number of Atoms in Compound	Mass of Element in Compound
C	12.0	3	36.0
H	1.0	6	6.0
O	16.0	1	16.0
		Formula mass =	58.0

Use Equation *4* (percent composition) on Reference Table *T*. Since the contribution of oxygen to the compound's mass is 16, we can calculate the percent composition by mass as follows:

$$\frac{16.0}{58.0} \times 100\% = \textbf{27.6\%}$$

9. **4** The sharing of one or more pairs of electrons by two atoms is known as a covalent bond.

Wrong Choices Explained:
    (1) The transfer of electrons forms an *ionic bond*.
    (2) Electron capture by a nucleus is a type of *nuclear reaction*.
    (3) The existence of mobile electrons within a metal is the basis of the *metallic bond*.

10. **3** Refer to Reference Table *S*. The respective electronegativities of carbon (C) and fluorine (F) are 2.6 and 4.0. Therefore, each of the carbon-fluorine bonds are polar. However, the CF_4 molecule has a tetrahedral shape and is entirely symmetrical. As a result, the entire molecule is nonpolar.

11. **3** When any atom loses electrons, it becomes a positive ion. Refer to the Periodic Table of the Elements. The electron configuration of an atom of barium is 2-8-18-18-8-2. The electron configuration of a Ba^{2+} ion is 2-8-18-18-8. The loss of the sixth shell reduces the size of the ion. Therefore, the radius of Ba^{2+} is smaller than the radius of Ba.

12. **4** The characteristic of a metallic bond is the presence of mobile valence electrons. This mobility accounts for the conductivity of metallic substances.

13. **3** The attraction of an atom for electrons in a chemical bond is measured by the electronegativity of the atom. Refer to Reference Table S. Of the choices given, choice (3), nitrogen, has the smallest electronegativity, 3.0.

14. **°1** Salt is nonvolatile, and evaporating the water would leave the salt behind.

> °Since the question did not make clear whether the salt was *completely* dissolved in the water, credit is also allowed for choice (2), *filtration*.

15. **2** The average molecular kinetic energy of a collection of molecules depends solely on the *temperature* of the substance: the higher the temperature, the greater the average molecular kinetic energy.

16. **3** A gas behaves most ideally when intermolecular attractions are at their lowest. At *low pressures*, the molecules are farther apart and are less able to form significant intermolecular attractions. At *high temperatures*, the greater average kinetic energy (that is, speed) of the molecules prevents the molecules from forming significant intermolecular attractions.

17. **4** Since the bonding in O_2 and O_3 are different, these forms of oxygen behave differently, both physically and chemically.

18. **4** Particles arranged in a rigid, fixed, geometric pattern are characteristic of the *solid* phase.

19. **3** Zinc powder has a much larger surface area than a strip of zinc. As a result, many more atoms of zinc are exposed to the HCl when the zinc is in the powdered form.

Wrong Choice Explained:
(4) The powdered zinc has the *same number of atoms* as the zinc strip since both have a mass of 5.0 grams. However, the powder has more *exposed atoms* than the strip.

20. **3** In chemistry, equilibrium is defined as the state that exists when the forward and reverse reaction rates are equal.

21. **1** The activation energy is the minimum energy needed by the reactants to begin a chemical reaction. A catalyst provides an alternative path with a lower activation energy for the reaction. As a result, more atoms can react in a given amount of time. In other words, the rate of the reaction increases.

22. **3** All organic compounds must contain carbon.

23. **1** A hydrocarbon contains only carbon and hydrogen. A saturated organic compound contains only carbon-carbon single bonds. The name of a saturated hydrocarbon will always end in -ane. Hexane is a saturated hydrocarbon containing six carbon atoms.

Wrong Choices Explained:
(2) Hexene is a hydrocarbon that contains six carbon atoms and one carbon-carbon double bond.
(3) Hexanol is an alcohol, which contains oxygen in addition to carbon and hydrogen.
(4) Hexanal is an aldehyde, which contains oxygen in addition to carbon and hydrogen.

24. **4** In this reaction, an organic acid and an alcohol react to produce an *ester* and water.

Wrong Choices Explained:
(1) Fermentation is the oxidation of a sugar to produce an alcohol and carbon dioxide.
(2) Saponification is the reaction of a fat (an ester) with a base to produce soap.
(3) Hydrogenation is an addition reaction used to saturate organic compounds that are unsaturated.

25. **1** Refer to Reference Table R. The properties of ethanoic acid, an organic acid, are due to the presence of the -COOH (carboxyl) functional group. Of the choices given, only choice (1), C_3H_7COOH, contains the carboxyl functional group.

Wrong Choices Explained:
(2), (3), (4) These compounds are, respectively, an alcohol, an ester, and an ether. Each of these compounds has properties distinctly different from an organic acid.

26. **1** As a free element, Zn(s) has an oxidation number of 0. In a compound such as $ZnSO_4$, the *sum* of the oxidation numbers must add to 0. Refer to Reference Table *E*. Since the SO_4^{2-} ion has an oxidation number of –2 (its charge), the Zn in $ZnSO_4$ must have an oxidation number of +2. Therefore, the oxidation number of Zn changes from 0 to +2.

27. **2** A voltaic cell is a type of electrochemical cell that generates electrical energy by converting the chemical energy obtained from a spontaneous oxidation-reduction reaction.

28. **2** An electrolyte is a substance that conducts electricity when dissolved in water. The conductivity is due to the presence of mobile ions in the solution. HCl conducts electricity in aqueous solution because it separates into H^+ and Cl^- ions. NaOH is an ionic compound. In aqueous solution, the Na^+ and OH^- ions separate and are able to conduct electricity.

Wrong Choices Explained:
(1), (3), (4) CH_3OH dissolves in water, but it does not separate into ions. C_5H_{12} is a nonpolar substance that does not dissolve in water.

29. **1** An Arrhenius acid is defined as a substance that produces H^+ ions in aqueous solution.

30. **2** In a neutralization reaction, an acid and a hydroxide base combine to form a salt and water. Of the choices given, choice (2), KOH, is a hydroxide base. None of the other choices represent an acid or a base.

31. **1** Refer to Reference Table *O*. An alpha particle is the nucleus of a helium-4 atom and contains 2 protons and 2 neutrons.

Wrong Choices Explained:
(2), (4) A beta particle and a positron are, respectively, negative and positive electrons, particles with approximately 2,000 times *less* mass than a proton.

32. **1** The energy produced in a nuclear fusion reaction is produced because some of the mass is converted into energy. As a result, the mass of the products is less than the mass of the reactants.

33. **3** Penetrating power is the ability of radiation to pass through material substances. The penetrating power of the particles in this question, in decreasing order, is:

$$\text{gamma} > \text{beta} \approx \text{positron} > \text{alpha}$$

PART B–1

34. **1** Refer to the Periodic Table of the Elements. The atomic number of Au is 79. An atom of Au contains 79 electrons. When an Au^{3+} ion is formed, 3 electrons are lost. Therefore, Au^{3+} contains 76 electrons.

35. **3** Refer to the Periodic Table of the Elements. All of the electron configurations shown in it are *ground-state* configurations, that is, electrons in their lowest energy levels. When an electron is "promoted" to a higher level, the atom is said to be excited. Choice (3), 2-7-2, represents an excited Na atom. In the ground state, the configuration of Na is 2-8-1. In the 2-7-2 configuration, one of the electrons in the second level has been raised to the third level.

Wrong Choices Explained:
(1), (2), (4)   These choices represent the ground-state configurations of C, O, and Mg, respectively.

36. **2** In the general atomic symbol $^{A}_{Z}X$, Z is the atomic number, the number of protons in the nucleus of the atom; A is the mass number, the *sum* of the protons and neutrons in the nucleus. The expression $A - Z$ equals the number of neutrons in the nucleus. The atom $^{19}_{9}F$ contains 9 protons (and 9 electrons) and 10 neutrons. The atom $^{12}_{6}C$ contains 6 protons (and 6 electrons) and 6 neutrons. Therefore, $^{12}_{6}C$ has *three fewer* protons (and electrons) and *four fewer* neutrons than $^{19}_{9}F$. This eliminates choices (1), (3), and (4). Refer to the Periodic Table of the Elements. The ground-state configuration of $^{12}_{6}C$ is 2-4, while the ground-state configuration of $^{19}_{9}F$ is 2-7. Therefore, $^{12}_{6}C$ contains *three* fewer *valence* electrons than $^{19}_{9}F$.

37. **4** The description that X is brittle and lacks luster indicates that this element is a nonmetal. As such, it should be found on the right side of the Periodic Table of the Elements. In addition, the presence of six valence electrons indicates that X is located in Group 16.

38. **1** An empirical formula is one whose atoms are expressed in *smallest* whole-number ratios. To obtain the empirical formula of $C_6H_{12}O_6$, divide the formula by 6: CH_2O.

39. **1** Refer to Reference Table S. The respective electronegativities of hydrogen (H) and oxygen (O) are 2.1 and 3.4. Therefore, each of the hydrogen-oxygen bonds are polar covalent.

40. **2** Since heat is continually absorbed by the substance, the potential energy and/or the average kinetic energy continually increase. When the temperature of the substance remains constant, as during a phase change, the potential energy increases <u>but</u> the <u>ave</u>rage kinetic energy does *not* increase. This occurs over segments *BC* and *DE*.

41. **2** The respective vapor pressures of propanone and water at 50°C are 74 kPa and 12 kPa. The vapor pressure of a substance measures the ability of molecules to escape from the liquid to the gas phase. Propanone molecules escape more easily than water molecules because the intermolecular forces among propanone molecules are *weaker* than the intermolecular forces among water molecules.

42. **3** When a solution is in equilibrium, the rate at which excess solute dissolves must equal the rate at which the dissolved solute leaves the solution. Since these two rates are equal, dissolving additional solute at a given temperature and pressure is not possible. In other words, the solution is *saturated*.

43. **2** Increasing the concentration of $N_2(g)$ would increase the number of effective collisions between $N_2(g)$ and $O_2(g)$, leading to an increase in the rate of the forward reaction.

Wrong Choices Explained:
(1) Increasing the concentration of $NO(g)$ would lead to an increase in the rate of the *reverse* reaction.
(3) Decreasing the temperature always decreases the rates of both the forward and the reverse reactions.
(4) Decreasing the reaction pressure effectively decreases the concentrations of (gaseous) reactants and products. As a result, the rates of the forward and reverse reactions would decrease.

44. **1** A substance in the gas phase always has a greater degree of disorder than the same substance in either the solid or liquid phases. If two samples of a substance are at different temperatures, the sample at the higher temperature has the greater disorder.

45. **4** The pH scale effectively runs from 0–14. Any pH greater than 7 indicates a basic solution; any pH less than 7 indicates an acidic solution. A pH of 7 indicates a neutral solution.

46. **4** Refer to Reference Tables P and Q. The prefix pent- indicates the presence of *five carbon atoms*. The ending -yne indicates the presence of one carbon-carbon *triple bond*. Of the choices given, only choice (4) meets both of these requirements.

47. **4** Refer to Reference Table R. An ether has the general formula R–O–R', where R and R' represent a bonded carbon atom or a group of carbon atoms. The structural formula given in choice (4) represents dimethyl ether.

Wrong Choices Explained:
(1) The structural formula given in this choice represents the *aldehyde* ethanal.
(2) The structural formula given in this choice represents the *alcohol* ethanol.
(3) The structural formula given in this choice represents the *organic acid* ethanoic acid.

48. **2** As a free element, Al has an oxidation number of 0. Refer to the Periodic Table of the Elements. When in Al_2O_3, the Al has an oxidation number of +3 since it exists as the Al^{3+} ion. Al forms a 3+ ion by *losing* three electrons. The correct oxidation half-reaction is:

$$Al \rightarrow Al^{3+} + 3e^-$$

49. **2** Gold-198 has a half-life of 2.69 days. After this period of time has elapsed, $\frac{1}{2}$ of the gold will remain unchanged.

50. **1** Refer to Reference Table S and to the Periodic Table of the Elements. The elements of Group 1, in order of increasing atomic radius, are Li, Na, K, Rb, Cs, and Fr. (H does not really belong in Group 1.) Inspection of Table S indicates that as the atomic radius increases, the first ionization energy generally decreases.

PART B–2

[Point values are indicated in brackets.]

51. In this *decomposition* reaction, water is separated into hydrogen and oxygen. Other acceptable examples include analysis, redox, endothermic, and electrolysis. [1 point]

52. The same number of atoms of each element appear on both sides of the equation: 4 hydrogen atoms and 2 oxygen atoms.   [1 point]

53. Refer to the coefficients of this equation. 2 moles of H_2O produce 1 mole of O_2:

$$8 \text{ mol } H_2O \bullet \left(\frac{1 \text{ mol } O_2}{2 \text{ mol } H_2O} \right) = \mathbf{4 \text{ mol } H_2O}$$

[1 point]

54. In this reaction, the Ag^+ ion is reduced to Ag:

$$Ag^+ + e^- \rightarrow Ag$$

$$2Ag^+ + 2e^- \rightarrow 2Ag$$

Note that either equation is acceptable for credit.   [1 point]

55. The balanced equation is:

$$Cu(s) + \mathbf{2}AgNO_3(aq) \rightarrow Cu(NO_3)_2(aq) + \mathbf{2}Ag(s)$$

Note that also placing a "1" before $Cu(s)$ and/or $Cu(NO_3)_2(aq)$ is acceptable for credit.   [1 point]

56. Refer to Reference Table *M*. Phenolphthalein changes from *colorless to pink*.   [1 point]

57. Use Equation 7 on Reference Table *T*:

$$M_A V_A = M_B V_B$$
$$M_A = \frac{M_B V_B}{V_A}$$

$$= \frac{\mathbf{(0.30 \text{ M}) \bullet (42.2 \text{ mL})}}{\mathbf{(60.0 \text{ mL})}}$$

[1 point]

58. Since the molarity of the base is expressed to 2 significant figures, the answer must also be expressed to 2 significant figures.   [1 point]

59. Isotopes differ in the *number of neutrons* present in the nucleus. [1 point]

60. The atomic mass of an element is determined by calculating the *weighted average* of the masses of each isotope. Multiply the mass of each isotope by the *decimal equivalent* of its percent abundance, and then add the products together:

$$(0.909) \bullet (19.99) + (0.003) \bullet (20.99) + (0.088) \bullet (21.99)$$

[1 point]

61. Since neon-20 has the greatest abundance, the atomic mass of neon will be closest to the isotopic mass of this isotope, *20*.   [1 point]

62. Na and K have similar chemical properties because both *elements are located in the same group* (Group 1).

Other acceptable answers include:

- They have the same number of valence electrons.
- Both elements form 1+ ions.
- Both elements are alkali metals.

[1 point]

63. The equation given in the answer booklet is shown below:

This equation represents an *addition reaction*, in which a molecule of Br_2 is "added" across the double bond in the hydrocarbon (2-butene). One bromine atom is bonded to carbon atom 2, and one bromine atom is bonded to carbon atom 3. As a result, the double bond between carbon atoms 2 and 3 is reduced to a single bond. Any of the structures shown below is acceptable for credit:

$$
\begin{array}{cccc}
H & H & H & H \\
| & | & | & | \\
H-C-&C-&C-&C-H \\
| & | & | & | \\
H & Br & Br & H
\end{array}
$$

$$
\begin{array}{cccc}
 & Br & & \\
| & | & | & | \\
-C-&C-&C-&C- \\
| & | & | & | \\
 & Br & &
\end{array}
$$

$$CH_3-CH-CH-CH_3$$
$$\qquad \;\; | \qquad \; |$$
$$\qquad \;\; Br \quad \; Br$$

$$CH_3CHBr\,CHBr\,CH_3 \qquad\qquad \text{[1 point]}$$

64. In order for a nuclear equation to be balanced, the atomic numbers (nuclear charges) and the mass numbers on both sides of the equation must be equal. By applying this rule, we see that X has a nuclear charge of +1 and a mass number of 0. Refer to Reference Table O. The particle is identified as a *positron*.

Note that $_{+1}^{0}e$, $_{+1}^{0}\beta$, or β^+ are also acceptable for credit.   [1 point]

PART C

[Point values are indicated in brackets.]

65. The expression shown below can be used to calculate the number of moles in the sample:

$$8.40 \text{ g} \cdot \left(\frac{1 \text{ mol}}{58.3 \text{ g}}\right)$$

[1 point]

66. According to the equation, 2 moles of HCl are needed to neutralize 1 mole of $Mg(OH)_2$:

$$0.05 \text{ mol HCl} \cdot \left(\frac{1 \text{ mol } Mg(OH)_2}{2 \text{ mol HCl}}\right) = \mathbf{0.025 \text{ mol } Mg(OH)_2}$$

[1 point]

67. Use Reference Table S to obtain the symbol for sodium (Na) and Reference Table E to obtain the formula and charge for the polyatomic ion hydrogen carbonate (HCO_3^-). Since sodium forms an ion whose charge is 1+, the formula for sodium hydrogen carbonate is *NaHCO$_3$*.   [1 point]

68. A substance that sublimes changes directly from the solid to the gas phase. In order to exhibit this property, the substance must have *weak intermolecular forces*.   [1 point]

69. Water is a polar solvent and will dissolve substances that are also polar. Since naphthalene is nonpolar, it will not dissolve appreciably in water. Chemists frequently use the expression "like dissolves like."   [1 point]

70. In order to determine the molecular formula of naphthalene:

- Calculate the gram-formula mass of the *empirical formula*:

$$(5) \cdot (12) + (4) \cdot (1) = 64$$

- Divide this number into the gram-formula mass of naphthalene:

$$\frac{128}{64} = 2$$

- Multiply the empirical formula by the number you obtained in the previous step.

$$2 \cdot (C_5H_4) = \mathbf{C_{10}H_8}$$

[1 point]

71. One appropriate scale is to mark the y-axis from 0 to 50. In other words, every four boxes equals 10 grams per 100 grams of H_2O. This scale is shown in the diagram following the answer to question 72.   [1 point]

72. One credit is allowed for plotting all points correctly (within ±0.3 grid space). The diagram below shows the completed graph:

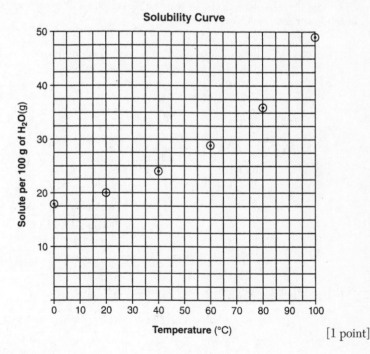

Solubility Curve

[1 point]

73. According to the data table, a saturated solution contains 24 grams of solute per 100 grams of water at 40°C. Since only 15 grams of solute have been dissolved, 9 *additional grams* of solute are needed to saturate the solution at this temperature.   [1 point]

74. According to the solubility curve for $KClO_3$ found on Reference Table G, the solubility of this compound is approximately 7 grams per 100 grams of water. Therefore 7 *grams of* $KClO_3$ must be added to the water.

Note that any mass in the range of 6–8 grams is acceptable for credit. [1 point]

75. Decreasing the temperature reduces the average kinetic energy of the gas particles in the balloon. In other words, the particles will move more slowly at the lower temperature.   [1 point]

76. Use Equation 6 (combined gas law) on Reference Table T:

$$\frac{P_1 V_1}{T_1} = \frac{P_2 V_2}{T_2}$$

$$\frac{(100.8 \text{ kPa}) \bullet (52.5 \text{ L})}{(295 \text{ K})} = \frac{(45.6 \text{ kPa}) \bullet (V_2)}{(252 \text{ K})}$$

$$V_2 = \frac{(100.8 \text{ kPa}) \bullet (52.5 \text{ L}) \bullet (252 \text{ K})}{(295 \text{ K}) \bullet (45.6 \text{ kPa})}$$

[1 point]

77. Use Equation 9 (temperature) on Reference Table T:

$$K = {}^\circ C + 273$$
$$^\circ C = K - 273$$
$$= 252 - 273$$
$$= -21^\circ C$$

[1 point]

78. Use the information given on Reference Table A:

$$45.6 \text{ kPa} \bullet \left(\frac{1 \text{ atm}}{101.3 \text{ kPa}}\right) = \textbf{0.450 atm}$$

[1 point]

79. When CO_2 is removed from the system, the state of equilibrium is disturbed. As a response to this disturbance, the system *shifts to the right* in order to replace some of the CO_2. As a result, the concentration of H_2CO_3 decreases.   [1 point]

80. In combination, both H atoms have an oxidation number of +1 and all three O atoms have an oxidation number of –2. The sum of all the oxidation numbers in this *neutral compound* must add to 0:

$$(2) \bullet (+1) + (3) \bullet (-2) + X = 0$$
$$X = \textbf{+4}$$

[1 point]

81. The fluoride ion has a charge of 1– and has 8 valence electrons:

Note that credit is awarded even if the brackets are omitted.   [1 point]

82. The designation "tin(II)" means that tin has a charge of 2+. The formula for tin(II) fluoride is: SnF_2.   [1 point]

83. Use Equation 5 (concentration—parts per million) on Reference Table T:

$$\text{parts per million} = \frac{\text{grams of solute}}{\text{grams of solution}} \times 1,000,000$$

$$= \frac{0.000250 \text{ g}}{175 \text{ g}} \times 1,000,000$$

$$= \textbf{1.43 ppm}$$

[1 point]

84. According to the passage, the Environmental Protection Agency has established a fluoride ion limit of 4 ppm. Since 1.43 ppm is less than this limit, the town's water is safe to drink. [1 point]

85. A number of precautions can be used as acceptable answers:

- The test should be performed under a fume hood.
- Avoid spilling the acid.
- In case of an acid spill, spread powdered $NaHCO_3$ over the spill.
- Avoid wearing clothing, such as open-toed shoes, that would not prevent spilled acid from penetrating the skin. [1 point]

Mark (✓) the questions you answered correctly. Count the number of checks and follow the formulas given to determine your score on each topic.

Core Area	☐ Questions Answered Correctly
	8, 71, 72, 73, 77, 78, 84
Section M—Math Skills ☐ Number of checks ÷ 7 × 100 = _____%	
	1, 2, 3, 31, 34, 35, 36, 60, 61
Section I—Atomic Concepts ☐ Number of checks ÷ 9 × 100 = _____%	
	4, 5, 6, 37, 50, 59, 62
Section II—Periodic Table ☐ Number of checks ÷ 7 × 100 = _____%	
	7, 8, 38, 51, 52, 53, 54, 55, 65, 66, 67, 70, 82
Section III—Moles/Stoichiometry ☐ Number of checks ÷ 13 × 100 = _____%	
	9, 10, 11, 12, 13, 39, 81
Section IV—Chemical Bonding ☐ Number of checks ÷ 7 × 100 = _____%	
	14, 15, 16, 17, 18, 40, 41, 68, 69, 75, 76, 83
Section V—Physical Behavior of Matter ☐ Number of checks ÷ 12 × 100 = _____%	
	19, 20, 21, 42, 43, 44, 79
Section VI—Kinetics and Equilibrium . ☐ Number of checks ÷ 7 × 100 = _____%	
	22, 23, 24, 25, 46, 47, 63
Section VII—Organic Chemistry ☐ Number of checks ÷ 7 × 100 = _____%	
	26, 27, 48, 80, 85
Section VIII—Oxidation-Reduction ☐ Number of checks ÷ 5 × 100 = _____%	
	28, 29, 30, 45, 56, 57
Section IX—Acids, Bases, and Salts ☐ Number of checks ÷ 6 × 100 = _____%	
	31, 32, 33, 49, 64
Section X—Nuclear Chemistry ☐ Number of checks ÷ 5 × 100 = _____%	

Examination August 2004

Chemistry
The Physical Setting

PART A

Answer all questions in this part.

Directions (1–33): For *each* statement or question, write in the answer space the *number* of the word or expression that, of those given, best completes the statement or answers the question. Some questions may require the use of the *Reference Tables for Physical Setting/Chemistry*.

1 Which of these phrases best describes an atom?

(1) a positive nucleus surrounded by a hard negative shell

(2) a positive nucleus surrounded by a cloud of negative charges

(3) a hard sphere with positive particles uniformly embedded

(4) a hard sphere with negative particles uniformly embedded

1_____

2 Which statement is true about a proton and an electron?

  (1) They have the same masses and the same charges.

  (2) They have the same masses and different charges.

  (3) They have different masses and the same charges.

  (4) They have different masses and different charges.   2_____

3 The atomic mass of an element is the weighted average of the masses of

  (1) its two most abundant isotopes

  (2) its two least abundant isotopes

  (3) all of its naturally occurring isotopes

  (4) all of its radioactive isotopes   3_____

4 What determines the order of placement of the elements on the modern Periodic Table?

  (1) atomic number

  (2) atomic mass

  (3) the number of neutrons, only

  (4) the number of neutrons and protons   4_____

5 Which compound contains only covalent bonds?

  (1) $NaOH$ (3) $Ca(OH)_2$

  (2) $Ba(OH)_2$ (4) CH_3OH 5_____

6 At 298 K, oxygen (O_2) and ozone (O_3) have different properties because their

  (1) atoms have different atomic numbers

  (2) atoms have different atomic masses

  (3) molecules have different molecular structures

  (4) molecules have different average kinetic energies   6_____

7 Which substance represents a compound?

    (1) $C(s)$ (3) $CO(g)$

    (2) $Co(s)$ (4) $O_2(g)$ 7_____

8 All chemical reactions have a conservation of

    (1) mass, only

    (2) mass and charge, only

    (3) charge and energy, only

    (4) mass, charge, and energy   8_____

9 Which characteristic is a property of molecular substances?

    (1) good heat conductivity

    (2) good electrical conductivity

    (3) low melting point

    (4) high melting point   9_____

10 Given the Lewis electron-dot diagram:

$$\begin{array}{c} H \\ \cdot\cdot \\ H:C:H \\ \cdot\cdot \\ H \end{array}$$

Which electrons are represented by all of the dots?

    (1) the carbon valence electrons, only

    (2) the hydrogen valence electrons, only

    (3) the carbon and hydrogen valence electrons

    (4) all of the carbon and hydrogen electrons   10_____

11 Which grouping of the three phases of bromine is listed in order from left to right for increasing distance between bromine molecules?

    (1) gas, liquid, solid   (3) solid, gas, liquid

    (2) liquid, solid, gas   (4) solid, liquid, gas   11_____

12 Which statement concerning elements is true?

(1) Different elements must have different numbers of isotopes.

(2) Different elements must have different numbers of neutrons.

(3) All atoms of a given element must have the same mass number.

(4) All atoms of a given element must have the same atomic number.

12_____

13 At room temperature, the solubility of which solute in water would be most affected by a change in pressure?

(1) methanol   (3) carbon dioxide

(2) sugar   (4) sodium nitrate

13_____

14 Based on Reference Table *I*, which change occurs when pellets of solid NaOH are added to water and stirred?

(1) The water temperature increases as chemical energy is converted to heat energy.

(2) The water temperature increases as heat energy is stored as chemical energy.

(3) The water temperature decreases as chemical energy is converted to heat energy.

(4) The water temperature decreases as heat energy is stored as chemical energy.

14_____

15 The concept of an ideal gas is used to explain

(1) the mass of a gas sample

(2) the behavior of a gas sample

(3) why some gases are monatomic

(4) why some gases are diatomic

15_____

16 Molecules in a sample of $NH_3(\ell)$ are held closely together by intermolecular forces

  (1) existing between ions
  (2) existing between electrons
  (3) caused by different numbers of neutrons
  (4) caused by unequal charge distribution   16_____

17 Which process represents a chemical change?

  (1) melting of ice
  (2) corrosion of copper
  (3) evaporation of water
  (4) crystallization of sugar   17_____

18 At STP, which 4.0-gram zinc sample will react fastest with dilute hydrochloric acid?

  (1) lump   (3) powdered
  (2) bar   (4) sheet metal   18_____

19 Which information about a chemical reaction is provided by a potential energy diagram?

  (1) the oxidation states of the reactants and products
  (2) the average kinetic energy of the reactants and products
  (3) the change in solubility of the reacting substances
  (4) the energy released or absorbed during the reaction   19_____

20 A catalyst works by

  (1) increasing the potential energy of the reactants
  (2) increasing the energy released during a reaction
  (3) decreasing the potential energy of the products
  (4) decreasing the activation energy required for a reaction   20_____

21 Even though the process is endothermic, snow can sublime. Which tendency in nature accounts for this phase change?

(1) a tendency toward greater entropy
(2) a tendency toward greater energy
(3) a tendency toward less entropy
(4) a tendency toward less energy   21_____

22 What is the IUPAC name of the compound with the structural formula shown below?

(1) 2-pentene   (3) 2-pentyne
(2) 3-pentene   (4) 3-pentyne   22_____

23 Molecules of 1-bromopropane and 2-bromo-propane differ in

(1) molecular formula
(2) structural formula
(3) number of carbon atoms per molecule
(4) number of bromine atoms per molecule   23_____

24 Which half-reaction correctly represents reduction?

(1) $Ag \rightarrow Ag^+ + e^-$ (3) $Au^{3+} + 3e^- \rightarrow Au$
(2) $F_2 \rightarrow 2 F^- + 2e^-$ (4) $Fe^{2+} + e^- \rightarrow Fe^{3+}$ 24_____

25 In a redox reaction, how does the total number of electrons lost by the oxidized substance compare to the total number of electrons gained by the reduced substance?

(1) The number lost is always greater than the number gained.
(2) The number lost is always equal to the number gained.
(3) The number lost is sometimes equal to the number gained.
(4) The number lost is sometimes less than the number gained.

25_____

26 Which reaction is an example of an oxidation-reduction reaction?

(1) $AgNO_3 + KI \rightarrow AgI + KNO_3$
(2) $Cu + 2 AgNO_3 \rightarrow Cu(NO_3)_2 + 2 Ag$
(3) $2 KOH + H_2SO_4 \rightarrow K_2SO_4 + 2 H_2O$
(4) $Ba(OH)_2 + 2 HCl \rightarrow BaCl_2 + 2 H_2O$

26_____

27 Which compound is an Arrhenius base?

(1) CH_3OH (3) $LiOH$
(2) CO_2 (4) NO_2

27_____

28 The only positive ion found in an aqueous solution of sulfuric acid is the

(1) hydroxide ion   (3) sulfite ion
(2) hydronium ion   (4) sulfate ion

28_____

29 Which process uses a volume of solution of known concentration to determine the concentration of another solution?

(1) distillation   (3) transmutation
(2) substitution   (4) titration

29_____

30 Which pH change represents a hundredfold increase in the concentration of H_3O^+?

(1) pH 5 to pH 7   (3) pH 3 to pH 1

(2) pH 13 to pH 14   (4) pH 4 to pH 3   30_____

31 Which radioisotope undergoes beta decay and has a half-life of less than 1 minute?

(1) Fr-220   (3) N-16

(2) K-42   (4) P-32   31_____

32 Which set of symbols represents atoms with valence electrons in the same electron shell?

(1) Ba, Br, Bi   (3) O, S, Te

(2) Sr, Sn, I   (4) Mn, Hg, Cu   32_____

Note that question 33 has only three choices.

33 When compared with the energy of an electron in the first shell of a carbon atom, the energy of an electron in the second shell of a carbon atom is

(1) less

(2) greater

(3) the same   33_____

PART B-1

Answer all questions in this part.

Directions (34–50): For *each* statement or question, write in the answer space the *number* of the word or expression that, of those given, best completes the statement or answers the question. Some questions may require the use of the *Reference Tables for Physical Setting/Chemistry*.

34 What is the total number of electrons found in an atom of sulfur?

   (1) 6   (3) 16

   (2) 8   (4) 32   34_____

35 Which electron configuration represents the electrons of an atom in an excited state?

   (1) 2–8–1   (3) 2–8–17–6

   (2) 2–8–6   (4) 2–8–18–5   35_____

36 The nucleus of an atom of cobalt-58 contains

   (1) 27 protons and 31 neutrons

   (2) 27 protons and 32 neutrons

   (3) 59 protons and 60 neutrons

   (4) 60 protons and 60 neutrons   36_____

37 Which pair of formulas correctly represents a molecular formula and its corresponding empirical formula?

   (1) C_2H_2 and CH   (3) C_4H_6 and CH

   (2) C_3H_4 and CH_2 (4) C_5H_8 and C_2H_2 37_____

38 Which substance is correctly paired with its type of bonding?

(1) NaBr—nonpolar covalent
(2) HCl—nonpolar covalent
(3) NH_3—polar covalent
(4) Br_2—polar covalent   38_____

39 A gas occupies a volume of 444 mL at 273 K and 79.0 kPa. What is the final kelvin temperature when the volume of the gas is changed to 1880 mL and the pressure is changed to 38.7 kPa?

(1) 31.5 K   (3) 566 K
(2) 292 K   (4) 2360 K   39_____

40 At STP, which of these substances is most soluble in H_2O?

(1) CCl_4 (3) HCl
(2) CO_2 (4) N_2 40_____

41 Based on intermolecular forces, which of these substances would have the highest boiling point?

(1) He   (3) CH_4
(2) O_2 (4) NH_3 41_____

42 How much heat energy must be absorbed to completely melt 35.0 grams of $H_2O(s)$ at 0°C?

(1) 9.54 J   (3) 11 700 J
(2) 146 J   (4) 79 100 J   42_____

43 The graph below represents the uniform heating of a substance, starting below its melting point, when the substance is solid.

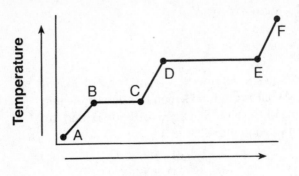

Time

Which line segments represent an increase in average kinetic energy?

(1) $\overline{AB}$ and $\overline{BC}$

(2) $\overline{AB}$ and $\overline{CD}$

(3) $\overline{BC}$ and $\overline{DE}$

(4) $\overline{DE}$ and $\overline{EF}$

43_____

44 Given the three organic structural formulas shown below:

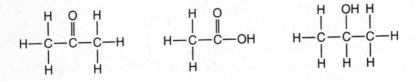

Which organic-compound classes are represented by these structural formulas, as shown from left to right?

(1) ester, organic acid, ketone

(2) ester, aldehyde, organic acid

(3) ketone, aldehyde, alcohol

(4) ketone, organic acid, alcohol

44_____

45 Given the reaction at equilibrium:

$$N_2(g) + O_2(g) + energy \rightleftharpoons 2\ NO(g)$$

Which change will result in a *decrease* in the amount of NO(g) formed?

(1) decreasing the pressure
(2) decreasing the concentration of $N_2(g)$
(3) increasing the concentration of $O_2(g)$
(4) increasing the temperature   45_____

46 Given the equation:

$$X + Cl_2 \rightarrow C_2H_5Cl + HCl$$

Which molecule is represented by X?

(1) C_2H_4 (3) C_3H_6
(2) C_2H_6 (4) C_3H_8 46_____

47 Which metal reacts spontaneously with a solution containing zinc ions?

(1) magnesium   (3) copper
(2) nickel   (4) silver   47_____

48 Which statement correctly describes a solution with a pH of 9?

(1) It has a higher concentration of H_3O^+ than OH^- and causes litmus to turn blue.
(2) It has a higher concentration of OH^- than H_3O^+ and causes litmus to turn blue.
(3) It has a higher concentration of H_3O^+ than OH^- and causes methyl orange to turn yellow.
(4) It has a higher concentration of OH^- than H_3O^+ and causes methyl orange to turn red.   48_____

49 How many days are required for 200. grams of radon-222 to decay to 50.0 grams?

(1) 1.91 days   (3) 7.64 days

(2) 3.82 days   (4) 11.5 days   49_____

50 A student calculates the density of an unknown solid. The mass is 10.04 grams, and the volume is 8.21 cubic centimeters. How many significant figures should appear in the final answer?

(1) 1   (3) 3

(2) 2   (4) 4   50_____

PART B–2

Answer all questions in this part.

Directions (51–65): Record your answers on the answer sheet provided in the back. Some questions may require the use of the *Reference Tables for Physical Setting/Chemistry*.

51 In the 19th century, Dmitri Mendeleev predicted the existence of a then unknown element X with a mass of 68. He also predicted that an oxide of X would have the formula X_2O_3. On the modern Periodic Table, what is the group number and period number of element X? [1]

52 Given the equation: $2 H_2(g) + O_2(g) \rightarrow 2 H_2O(g)$

If 8.0 moles of O_2 are completely consumed, what is the total number of moles of H_2O produced? [1]

53 In the space provided on the answer sheet, show a correct numerical setup for determining how many liters of a 1.2 M solution can be prepared with 0.50 mole of $C_6H_{12}O_6$. [1]

Base your answers to questions 54 through 57 on the particle diagrams below. Samples A, B, and C contain molecules at STP.

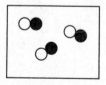
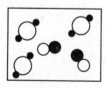
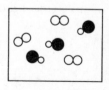

A   B   C

54 Explain why the average kinetic energy of sample *B* is equal to the average kinetic energy of sample *C*. [1]

55 Explain, in terms of the *composition,* why sample *A* represents a pure substance. [1]

56 Explain why sample *C* could represent a mixture of fluorine and hydrogen chloride. [1]

57 Contrast sample *A* and sample *B*, in terms of *compounds and mixtures*. Include both sample *A* and sample *B* in your answer. [1]

———

Base your answers to questions 58 through 60 on the electronegativity values and atomic numbers of fluorine, chlorine, bromine, and iodine that are listed on Reference Table *S*.

58 On the grid provided on your answer sheet, mark an appropriate scale on the axis labeled "Electronegativity." An appropriate scale is one that allows a trend to be seen. [1]

59 On the same grid, plot the electronegativity and atomic number data from Reference Table *S*. Circle and connect the points. [1]

Example: ○—○ ○

60 Explain, in terms of *electronegativity*, why the H–F bond is expected to be more polar than the H–I bond. [1]

_____

61 What is the gram-formula mass of $(NH_4)_2CO_3$? Use atomic masses rounded to the *nearest whole number.* [1]

62 In the space provided on your answer sheet, show a correct numerical setup for calculating the number of moles of CO_2 (gram-formula mass = 44 g/mol) present in 11 grams of CO_2. [1]

_____

Base your answers to questions 63 and 64 on the information below.

Given the equilibrium equation at 298 K:

$$KNO_3(s) + 34.89 \text{ kJ} \overset{H_2O}{\rightleftharpoons} K^+(aq) + NO_3^-(aq)$$

63 Describe, in terms of *LeChatelier's principle*, why an increase in temperature increases the solubility of KNO_3. [1]

64 The equation indicates that KNO_3 has formed a saturated solution. Explain, in terms of *equilibrium*, why the solution is saturated. [1]

_____

65 In the space provided on your answer sheet, draw the structural formula for butanoic acid. [1]

PART C

Answer all questions in this part.

Directions (66–85): Record your answers on the answer sheet provided in the back. Some questions may require the use of the *Reference Tables for Physical Setting/Chemistry.*

Base your answers to questions 66 through 69 on the information below, which describes the smelting of iron ore, and on your knowledge of chemistry.

In the smelting of iron ore, Fe_2O_3 is reduced in a blast furnace at high temperature by a reaction with carbon monoxide. Crushed limestone, $CaCO_3$, is also added to the mixture to remove impurities in the ore. The carbon monoxide is formed by the oxidation of carbon (coke), as shown in the reaction below:

$$2\,C + O_2 \rightarrow 2\,CO + energy$$

Liquid iron flows from the bottom of the blast furnace and is processed into different allays of iron.

66 Balance the equation for the reaction of Fe_2O_3 and CO on your answer sheet, using the smallest whole-number coefficients.   [1]

67 Using the set of axes provided on your answer sheet, sketch a potential energy diagram for the reaction of carbon and oxygen that produces carbon monoxide.   [1]

68 What is the oxidation number of carbon in $CaCO_3$?   [1]

69 Convert the melting point of iron metal to degrees Celsius. [1]

_____

Base your answers to questions 70 through 72 on the information below.

Potassium ions are essential to human health. The movement of dissolved potassium ions, $K^+(aq)$, in and out of a nerve cell allows that cell to transmit an electrical impulse.

70 What is the total number of electrons in a potassium ion? [1]

71 Explain, in terms of *atomic structure*, why a potassium ion is smaller than a potassium atom. [1]

72 What property of potassium ions allows them to transmit an electrical impulse? [1]

_____

Base your answers to questions 73 through 75 on the information below.

Ethene (common name ethylene) is a commercially important organic compound. Millions of tons of ethene are produced by the chemical industry each year. Ethene is used in the manufacture of synthetic fibers for carpeting and clothing, and it is widely used in making polyethylene. Low-density polyethylene can be stretched into a clear, thin film that is used for wrapping food products and consumer goods. High-density polyethylene is molded into bottles for milk and other liquids.

Ethene can also be oxidized to produce ethylene glycol, which is used in antifreeze for automobiles. The structural formula for ethylene glycol is:

```
      H   H
      |   |
  H — C — C — H
      |   |
      OH  OH
```

At standard atmospheric pressure, the boiling point of ethylene glycol is 198°C, compared to ethene that boils at −104°C.

73 Identify the type of organic reaction by which ethene (ehtylene) is made into polyethylene. [1]

74 According to the information in the reading passage, state *two* consumer products manufactured from ethene. [1]

75 Explain, in terms of *bonding*, why ethene is an unsaturated hydrocarbon. [1]

Base your answers to questions 76 through 78 on the diagram below, which represents a voltaic cell at 298 K and 1 atm.

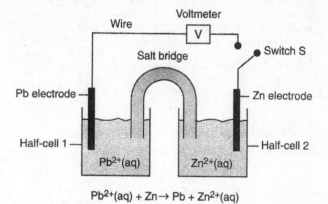

$$Pb^{2+}(aq) + Zn \rightarrow Pb + Zn^{2+}(aq)$$

76 In which half-cell will oxidation occur when switch *S* is closed?   [1]

77 Write the balanced half-reaction equation that will occur in half-cell 1 when switch *S* is closed.   [1]

78 Describe the direction of electron flow between the electrodes when switch *S* is closed.   [1]

Base your answers to questions 79 through 81 on the information and data table below.

Indigestion may be caused by excess stomach acid (hydrochloric acid). Some products used to treat indigestion contain magnesium hydroxide. The magnesium hydroxide neutralizes some of the stomach acid.

The amount of acid that can be neutralized by three different brands of antacids is shown in the data table on the next page.

Antacid Brand	Mass of Antacid Tablet (g)	Volume of HCl(aq) Neutralized (mL)
X	2.00	25.20
Y	1.20	18.65
Z	1.75	22.50

79 Based on Reference Table F, describe the solubility of magnesium hydroxide in water.   [1]

80 In the space provided on your answer sheet, show a correct numerical setup for calculating the milliliters of HCl(aq) neutralized per gram of antacid tablet for *each* brand of antacid.   [1]

81 Which antacid brand neutralizes the most acid per gram of antacid tablet?   [1]

———————————

Base your answers to questions 82 through 85 on the reading passage below and on your knowledge of chemistry.

A Glow in the Dark, and Scientific Peril

The [Marie and Pierre] Curies set out to study radioactivity in 1898. Their first accomplishment was to show that radioactivity was a property of atoms themselves. Scientifically, that was the most important of their findings, because it helped other researchers refine their understanding of atomic structure.

More famous was their discovery of polonium and radium. Radium was the most radioactive substance the Curies had encountered. Its radioactivity is due to the large size of the atom, which makes the nucleus unstable and prone to decay, usually to radon and then lead, by emitting particles and energy as it seeks a more stable configuration.

Marie Curie struggled to purify radium for medical uses, including early radiation treatment for tumors. But radium's bluish glow caught people's fancy, and companies in the United States began mining it and selling it as a novelty: for glow-in-the-dark light pulls, for instance, and bogus cure-all patent medicines that actually killed people.

What makes radium so dangerous is that it forms chemical bonds in the same way as calcium, and the body can mistake it for calcium and absorb it into the bones. Then, it can bombard cells with radiation at close range, which may cause bone tumors or bone-marrow damage that can give rise to anemia or leukemia.

—Denise Grady, *The New York Times*, October 6, 1998

82 State one risk associated with the use of radium. [1]

83 Using Reference Table *N*, complete the equation provided on your answer sheet for the nuclear decay of $^{226}_{88}$ Ra. Include *both* atomic number and mass number for *each* particle. [1]

84 Using information from the Periodic Table, explain why radium forms chemical bonds in the same way as calcium does. [1]

85 If a scientist purifies 1.0 gram of radium-226, how many years must pass before only 0.50 gram of the original radium-226 sample remains unchanged? [1]

Answer Sheet
August 2004
Chemistry
The Physical Setting

PART B–2

Answer all questions in Part B–2 and Part C. Record your answers on the answer sheet.

51 **Group** _____ and **Period** _____

52 _____ mol

53

54 _____

   _____

   _____

55 _____

   _____

   _____

56 _____

_____

_____

57 _____

_____

_____

58 and **59**

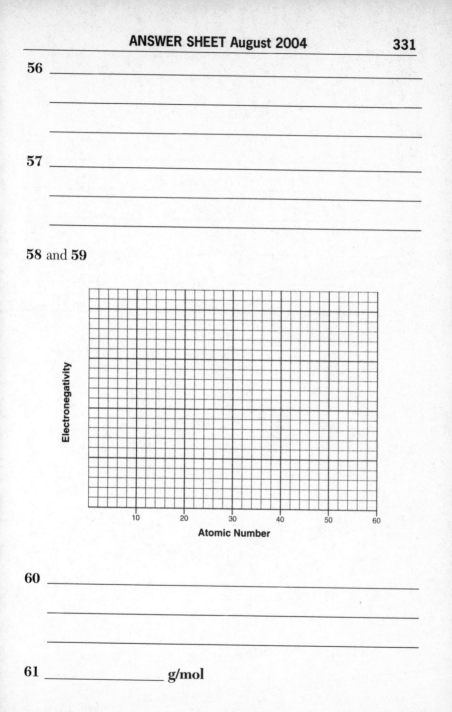

60 _____

_____

_____

61 _____ **g/mol**

62

63 _____

_____

_____

64 _____

_____

_____

65

PART C

66 _____Fe_2O_3 + _____CO → _____Fe + _____CO_2

67

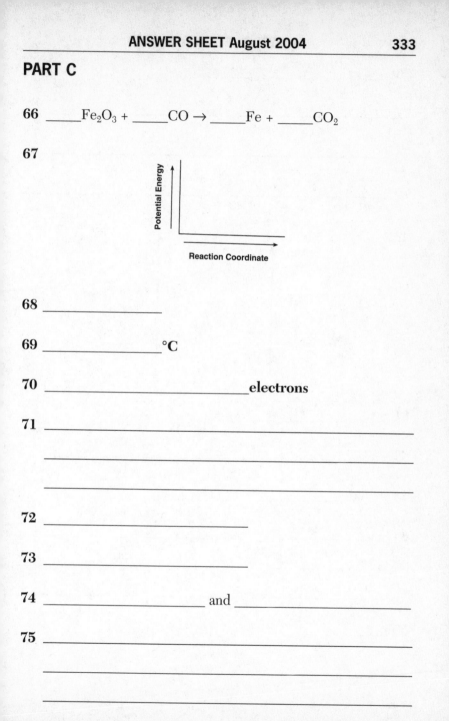

68 _____

69 _____ °C

70 _____ **electrons**

71 _____

_____

_____

72 _____

73 _____

74 _____ and _____

75 _____

_____

_____

76 _____

77 _____

78 _____

79 _____

80   *X*   *Y*   *Z*

81 _____

82 _____

83 $^{226}_{88}\text{Ra} \rightarrow$ _____ + _____

84 _____

_____

_____

85 _____y

Answers
August 2004

Chemistry
The Physical Setting

Answer Key

PART A

1. 2	**7.** 3	**13.** 3	**19.** 4	**25.** 2	**31.** 3
2. 4	**8.** 4	**14.** 1	**20.** 4	**26.** 2	**32.** 2
3. 3	**9.** 3	**15.** 2	**21.** 1	**27.** 3	**33.** 2
4. 1	**10.** 3	**16.** 4	**22.** 1	**28.** 2	
5. 4	**11.** 4	**17.** 2	**23.** 2	**29.** 4	
6. 3	**12.** 4	**18.** 3	**24.** 3	**30.** 3	

PART B–2

34. 3	**38.** 3	**42.** 3	**46.** 2	**50.** 3
35. 3	**39.** 3	**43.** 2	**47.** 1	
36. 1	**40.** 3	**44.** 4	**48.** 2	
37. 1	**41.** 4	**45.** 2	**49.** 3	

Answers Explained

PART A

1. **2** The current model of the atom consists of a positively charged nucleus, containing positively charged protons and neutral electrons, surrounded by a cloud of negatively charged electrons.

2. **4** An electron is the basic particle that carries a negative charge. A proton is the basic particle that carries a positive charge. The mass of a proton is nearly 2000 times greater than the mass of an electron. Therefore, a proton and an electron have different masses and different charges.

3. **3** By definition, the atomic mass of an element is the weighted average of the masses of all of its naturally occurring isotopes.

4. **1** Refer to the Periodic Table of the Elements. Elements are placed in order of increasing atomic number.

5. **4** The compound CH_3OH, known as methanol, is an organic compound in which all its atoms are bonded covalently.

Wrong Choices Explained:
(1), (2), (3) $NaOH$, $Ba(OH)_2$, and $Ca(OH)_2$ contain both ionic and covalent bonding. The metal ions (Na^+, Ba^{2+}, and Ca^{2+}) form ionic bonds with the negatively charged OH^- ion. Within the OH^- ion, the O and the H atoms are bonded covalently.

6. **3** Physical and chemical properties of substances are determined, in part, by their molecular structures. Since O_2 and O_3 have different structures, they have different chemical properties.

Wrong Choices Explained:
(1), (2) Since O_2 and O_3 contain only oxygen atoms, all of the atoms of oxygen in the two substances have the same atomic number and atomic mass.
(4) At the same temperature (298 K), the molecules of oxygen and ozone have the same average kinetic energy.

7. **3** A compound consists of two or more elements in chemical combination. The compound $CO(g)$ consists of molecules in which a carbon and an oxygen atom are bonded covalently.

Wrong Choices Explained:
   (1), (2), (4) $C(s)$, $Co(s)$, and $O_2(g)$ represent elements, not compounds.

8. **4** Three quantities are *always* conserved in all chemical reactions: mass, charge, and energy.

9. **3** Molecular substances have relatively weak *intermolecular forces*. As a result, they tend to be soft, have low melting and boiling points, and conduct electricity and heat poorly.

10. **3** In a Lewis electron-dot diagram, the *valence* (outer-level) electrons of the relevant atoms are represented by one or more pairs of dots.

11. **4** In most substances, atoms or molecules are closest together in the solid phase and farthest apart in the gas phase. Therefore, in bromine, the distance between the molecules increases from solid to liquid to gas.

12. **4** The atomic number—the number of protons in the nucleus of an atom—is the single quantity that identifies an element. Refer to the Periodic Table of the Elements and note that each element has its own unique atomic number.

13. **3** The solubility of gases is affected significantly by changes in pressure. The solubilities of solids and liquids are affected very little by changes in pressure. Of the choices given, only choice (3), carbon dioxide, is a gas.

Wrong Choices Explained:
   (1) Methanol is a liquid.
   (2), (4) Sugar and sodium nitrate are solids.

14. **1** According to Reference Table *I*, when NaOH dissolves in water, the reaction that occurs is:

$$NaOH(s) \xrightarrow{H_2O} Na^+(aq) + OH^-(aq)$$

In addition, 44.51 kilojoules of heat energy are released to the surroundings, increasing the temperature of the water. The source of the heat energy is the chemical energy stored in the reactants.

15. **2** In order to explain how a sample of a gas reacts to changes in pressure, volume, and temperature, the concept of the ideal gas was developed in the nineteenth century. An ideal gas is a collection of particles that have mass but negligible volume. The particles are in constant random motion and undergo perfect (elastic) collisions with other particles and the walls of the container. The model of the ideal gas is suitable for explaining the behavior of real gases, provided that the pressure is kept relatively low and the temperature is kept relatively high.

16. **4** Ammonia (NH_3) is a polar molecule. The unequal charge distribution within the molecule creates a negative end around the nitrogen atom and a positive end around the hydrogen atoms. In a sample of $NH_3(\ell)$, intermolecular attractions among the positive and negative ends of the molecules hold them closely together.

17. **2** A chemical change is one in which the identity or composition of one or more substances is changed. When copper corrodes, it combines with an element such as oxygen, and the element copper is changed into the compound copper oxide.

Wrong Choices Explained:
(1), (3), (4) Phase changes such as melting, evaporation, and crystallization do not change the identity or composition of a substance; these are *physical* changes.

18. **3** One of the factors that affect the rate of a reaction is the surface area of the reactants. Since powdered zinc has the greatest surface area, it will react most rapidly with the hydrochloric acid.

19. **4** Refer to the potential energy diagram shown below:

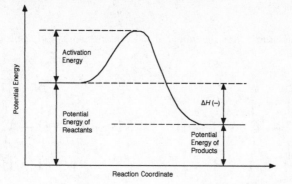

Of the choices given, only choice (4), the energy released or absorbed during the reaction, is provided by the potential energy diagram. (It is shown as ΔH.)

20. **4** A catalyst increases the rate of a reaction by providing an alternative pathway for the reaction to occur. (This is somewhat like certain roads in Europe that pass *through* mountains rather than go around them.) The alternative pathway has a lower activation energy than the original pathway. As a result, molecules with lower kinetic energies are capable of reacting, and the speed of the reaction is increased.

21. **1** Spontaneous processes depend on two factors: energy and entropy. Processes that are exothermic and lead to greater entropy will always be spontaneous. Nevertheless, it is possible for an endothermic reaction (such as sublimation) to occur spontaneously if the entropy increase and/or the temperature is sufficiently high.

22. **1** Refer to the diagram below and to Reference Tables *P* and *Q*:

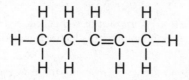

Note that the hydrocarbon has one double bond, classifying it as an alkene. Its name will end in -ene. Since the hydrocarbon has 5 carbon atoms, its name is pentene. Finally, note that the double bond is positioned between carbons 2

and 3. (*Count from right to left!*) Since the lower number is used, the IUPAC name is 2-pentene.

23. **2** Examine the diagrams of 1-bromopropane and 2-bromopropane given below:

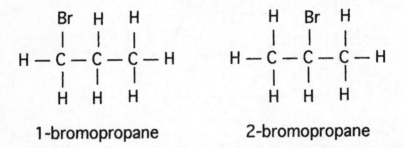

1-bromopropane   **2-bromopropane**

In 1-bromopropane, the bromine atom is bonded to the first carbon atom; in 2-bromopropane, the bromine atom is bonded to the middle carbon atom. Clearly, the *structures* of the molecules are different.

Wrong Choices Explained:
(1) Both compounds have the same molecular formula: C_3H_7Br.
(3) Both molecules contain 3 carbon atoms.
(4) Both molecules contain 1 bromine atom.

24. **3** Reduction represents the gain of electrons. In a reduction half-reaction, the product will have a lower oxidation number than that of the reactant, and the total charge will be the same on both sides of the reaction. Of the choices given, only choice (3), $Au^{3+} + 3e^- \rightarrow Au$, meets all three criteria.

Wrong Choices Explained:
(1) This half-reaction represents the *oxidation* of Ag to Ag^+.
(2), (4) These half-reactions are incorrectly written.

25. **2** Charge is *always conserved* in a chemical reaction. The number of electrons lost must equal the number of electrons gained.

26. **2** In an oxidation-reduction reaction, the transfer of electrons causes the oxidation number to change. The substance that is oxidized increases its oxidation number, while the substance that is reduced decreases its oxidation number. In choice (2), Cu is oxidized to Cu^{2+} and its oxidation number

increases from 0 to +2, while Ag^+ is reduced to Ag and its oxidation number decreases from +1 to 0.

Wrong Choices Explained:

(1), (3), (4) In each of these reactions, none of the elements undergoes a change in oxidation number. Therefore, they are not oxidation-reduction reactions.

27. **3** An Arrhenius base dissolves in water to produce OH^- ions as the sole negative ions in solution. See Reference Table F. LiOH is an ionic compound that is soluble in water; it dissolves to yield Li^+ and OH^- ions in solution.

Wrong Choices Explained:

(1) CH_3OH is an alcohol; the $-OH$ group is covalently bonded to the carbon atom. When dissolved in water, CH_3OH does not form OH^- ions in solution.

(2), (4) CO_2 and NO_2 are nonmetallic oxides. When dissolved in water, these gases actually produce *acidic* solutions.

28. **2** Sulfuric acid (H_2SO_4) is an Arrhenius acid: a substance that dissolves in water to produce hydronium (H_3O^+) ions as the sole positive ions in solution.

29. **4** The stem of this question defines titration: a process in which a volume of a solution of known concentration is used to determine the concentration of another solution.

30. **3** For each change of 1 pH unit, the H_3O^+ ion concentration changes by a factor of 10. As the pH *decreases*, the H_3O^+ ion concentration *increases*. In order for the H_3O^+ ion concentration to increase by 100 times, the pH must decrease by 2 pH units: from pH 3 to pH 1.

Wrong Choices Explained:

(1) This change represents a *hundredfold decrease* in the H_3O^+ ion concentration.

(2) This change represents a *tenfold decrease* in the H_3O^+ ion concentration.

(4) This change represents a *tenfold increase* in the H_3O^+ ion concentration.

31. **3** Refer to Reference Tables N and O. The radioisotope N-16 is a beta emitter with a half-life of 7.2 seconds.

Wrong Choices Explained:

(1) Fr-220 is an alpha emitter.

(2), (4) K-42 and P-32 are beta emitters, but their half-lives are longer than 1 minute.

32. **2** In order to have valence electrons in the same electron shell, elements must belong to the same period in the Periodic Table of the Elements. Of the choices given, only the elements in choice (2), Sr, Sn, I, belong to the same period (Period 5).

33. **2** Electrons in the second electron shell are, on the average, farther from the nucleus and contain more energy than electrons in the first electron shell.

PART B–1

34. **3** Refer to Reference Table S. The atomic number of sulfur is 16. An atom of sulfur contains 16 protons and 16 electrons.

35. **3** Refer to the Periodic Table of the Elements. All of the electron configurations given in the Periodic Table of the Elements are *ground-state* configurations, that is, electrons in their lowest energy levels. When an electron is promoted to a higher level, the atom is said to be excited. Choice (3), 2-8-17-6, represents an excited As atom. In the ground state, the configuration of As is 2-8-18-5. In the excited state, one of the electrons in the third level has been raised to the fourth level.

Wrong Choices Explained:
(1), (2), (4) These choices represent the ground-state configurations of Na, S, and As, respectively.

36. **1** In the atomic symbol, $_Z^A X$, Z is the atomic number, the number of protons in the nucleus of the atom; A is the mass number, the *sum* of the protons and neutrons in the nucleus. The expression $A - Z$ is equal to the number of neutrons in the nucleus. The atom cobalt-58, $_{27}^{58}Co$, contains 27 protons and (58 − 27) 31 neutrons.

37. **1** An empirical formula is one whose atoms are expressed in *smallest* whole number ratios. To obtain the empirical formula of C_2H_2, divide the formula by 2: CH.

38. **3** Refer to Reference Table S. The respective electronegativities of hydrogen (H) and nitrogen (N) are 2.1 and 3.0. Therefore, each of the hydrogen-nitrogen bonds are polar covalent.

Wrong Choices Explained:
(1) NaBr is ionic (the electronegativity difference is 2.1).
(2) HCl is polar covalent (the electronegativity difference is 1.1).
(4) Br_2 is nonpolar covalent (the electronegativity difference is 0.0).

39. **3** Use Equation 6 (the combined gas law) found on Reference Table T:

$$\frac{P_1 V_1}{T_1} = \frac{P_2 V_2}{T_2}$$

$$\frac{(79.0 \text{ kPa}) \cdot (444 \text{ mL})}{(273 \text{ K})} = \frac{(38.7 \text{ kPa}) \cdot (1880 \text{ mL})}{T_2}$$

$$T_2 = \textbf{566 K}$$

40. **3** Like dissolves like! H_2O is a polar substance and will dissolve ionic and polar substances best. The gas HCl is also polar and readily dissolves in H_2O.

Wrong Choices Explained:
(1), (2), (4) Each of these substances is nonpolar and will not dissolve well in H_2O.

41. **4** Intermolecular forces are strongest in polar covalent substances and weakest in nonpolar substances. Therefore, NH_3, a polar substance, will have the highest boiling point.

Wrong Choices Explained:
(1), (2), (3) Each of these substances is nonpolar and will have a relatively low boiling point.

42. **3** Refer to the heat of fusion given on Reference Table B. The heat of fusion is the quantity of heat energy needed to melt 1 gram of ice ($H_2O(s)$) at its melting point of 0°C:

$$(35.0 \text{ g}) \cdot \left(\frac{334 \text{ J}}{1 \text{ g}} \right) = \textbf{11700 J}$$

43. **2** Refer to the diagram associated with this question. Since temperature is a measure of average kinetic energy, an increase in average kinetic energy will occur whenever the temperature rises. This increase occurs over segments $\overline{AB}$, $\overline{CD}$, and $\overline{EF}$.

44. **4** Refer to Reference Table R, and compare the functional groups of the compounds with those given in the table. The first compound is a ketone, the second is an organic acid, and the third is an alcohol.

45. **2** When the concentration of $N_2(g)$ is decreased, the system will *shift to the left* in order to restore some of the removed $N_2(g)$. As a result, the concentration of $NO(g)$ will decrease.

Wrong Choices Explained:
(1) Since the equation has 2 moles of reactants and products, a change in pressure will leave the system unaffected.
(3) Increasing the $O_2(g)$ concentration will shift the reaction toward the right, leading to an *increase* in $NO(g)$.
(4) The forward reaction is endothermic. An increase in temperature favors the endothermic reaction, leading to an *increase* in $NO(g)$.

46. **2** The reaction given in this question represents a *substitution* reaction in which a Cl atom replaces an H atom. The full reaction is shown below:

$$\mathbf{C_2H_6} + Cl_2 \rightarrow C_2H_5Cl + HCl$$

47. **1** Refer to Reference Table J. Any metal that lies above zinc (Zn) will react spontaneously with Zn^{2+} ions. Of the choices given, only choice (1), magnesium (Mg), lies above Zn.

48. **2** Any pH above 7 represents a basic solution, that is, a solution in which the concentration of OH^- ions is greater than the concentration of H_3O^+ ions. Refer to Reference Table M. At a pH of 9, methyl orange will be yellow and litmus will be blue. Of the choices given, only choice (2) meets the criteria for ion concentration and for indicator color.

49. **3** Refer to Reference Table N. Radon-222 has a half-life of 3.82 days. 200. grams of this isotope will decay to 50.0 grams in 2 half-life periods ($200. \rightarrow 100. \rightarrow 50.0$) or 7.64 days.

50. **3** When dividing measured quantities, the answer should be expressed to the number of significant figures in the measurement with the *smallest* number of significant figures. Since 10.04 grams contains 4 significant figures and 8.21 cubic centimeters contains 3 significant figures, the answer should contain 3 significant figures.

PART B–2

[Point values are indicated in brackets.]

51. Refer to the Periodic Table of the Elements. Elements with an atomic mass of approximately 68 all lie in Period 4. The formula X_2O_3 indicates that X is an element that forms a 3+ ion. The elements of Group 13 all form 3+ ions. Therefore, X belongs in Group 13 and Period 4.   [1 point]

52. Refer to the coefficients of this equation. 1 mole of O_2 produces 2 moles of H_2O:

$$8.0 \text{ mol } O_2 \bullet \left(\frac{2 \text{ mol } H_2O}{1 \text{ mol } O_2} \right) = \textbf{16.0 mol } \textbf{H}_2\textbf{O}$$

[1 point]

53. Note that only the setup is required. Refer to Equation 5b (concentration equation) on Reference Table T:

$$\text{molarity } (M) \; = \frac{\text{moles of solute } (n)}{\text{liters of solution } (V)}$$

$$\textbf{1.2 M} \; = \frac{\textbf{0.50 mol}}{V}$$

[1 point]

54. Since both samples (B and C) are at the same temperature, they have equal average kinetic energies.   [1 point]

55. Sample A represents a pure substance because it contains only one type of particle.   [1 point]

56. Sample *C* represents a mixture of a diatomic element and a compound:

$$\text{OO} = F_2 \text{ and } \text{●} = HCl \qquad \text{[1 point]}$$

57. Sample *A* contains a pure substance consisting of two elements, a compound. Sample *C* contains a mixture of a compound and an element. [1 point]

58–59. Refer to the graph below. 1 point is allowed for marking an appropriate scale, and 1 point is allowed for plotting all the points correctly:

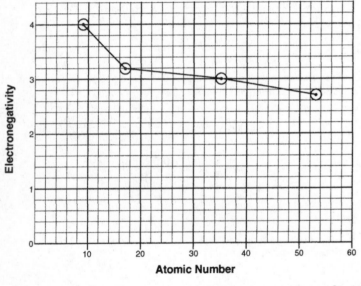

[2 points]

60. Bond polarity is determined by the difference in electronegativities. Since the electronegativity difference for H–F (1.9) is greater than the electronegativity difference of H–I (0.6), the H–F bond is more polar than the H–I bond. [1 point]

61. Complete the following table for $(NH_4)_2CO_3$, and add the numbers in the last column:

Element	Atomic Mass (g/mol)	Number of Atoms in Formula	Mass of Element in Formula/g
N	14	2	28
H	1	8	8
C	12	1	12
O	16	3	48
		Formula mass	**96**

[1 point]

62. Note that only the setup is required:

$$11 \text{ g } CO_2 \bullet \left(\frac{1 \text{ mol } CO_2}{44 \text{ g } CO_2} \right) \qquad \text{[1 point]}$$

63. The forward reaction (the dissolving of KNO_3) is endothermic. According to LeChatelier's principle, an increase in temperature favors the endothermic reaction. Therefore, increasing the temperature increases the solubility of KNO_3. [1 point]

64. In a saturated solution, the rate of KNO_3 dissolving in water is equal to the rate of the recrystallizing of KNO_3. [1 point]

65. Refer to the two possible diagrams of butanoic acid shown below:

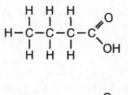

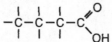

[1 point]

PART C

[Point values are indicated in brackets.]

66. In order to answer this question, the passage has to be read very carefully. According to the passage, Fe_2O_3 is *reduced* to Fe (see the first and last sentences). The reducing agent is carbon monoxide (CO). Therefore, the CO must be *oxidized* to CO_2. The balanced equation is:

$$(1)Fe_2O_3 + 3CO \rightarrow 2Fe + 3CO_2$$

Note that the coefficient "1" does not have to be included in order to receive credit.   [1 point]

67. The reaction between carbon and oxygen is exothermic. Refer to the potential energy diagram shown below:

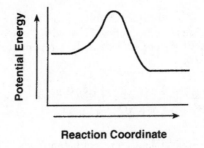

Reaction Coordinate   [1 point]

68. The sum of the oxidation numbers in a compound must equal zero. Since calcium is present as a 2+ ion, its oxidation number is +2. Each of the three oxygen atoms has an oxidation number of –2, accounting for a total of –6. In order for the sum to equal zero, the single carbon atom must have an oxidation number of **+4**.   [1 point]

69. Refer to Equation 9 on Reference Table *T*, and obtain the melting point of iron from Reference Table *S*:

$$K = {}^\circ C + 273$$
$${}^\circ C = K - 273 {}^\circ C = 1808 \text{ K} - 273 = \mathbf{1535 {}^\circ C} \qquad [1 \text{ point}]$$

70. See Reference Table S or refer to the Periodic Table of the Elements. The atomic number of potassium (K) is 19. An *atom* of potassium has 19 protons and 19 electrons. A potassium *ion*, K^+, has one less electron than the atom. Therefore, a K^+ ion contains **18 electrons**.   [1 point]

71. When an atom of potassium loses an electron to form a K^+ ion, the electron configuration changes from 2-8-8-1 to 2-8-8. The loss of an electron shell results in a smaller particle.   [1 point]

72. K^+ ions are charged particles. Moreover, when dissolved in water, they are free to move through the solution. Both of these factors contribute to the ability of K^+ ions to conduct electrical impulses.   [1 point]

73. Polyethylene is made from ethene by the process of **polymerization**. [1 point]

74. Consumer products manufactured from ethene include **synthetic fibers**, **clothing**, **carpeting**, **food wrap**, **bottles for containing liquids**, and **antifreeze**.

Note that you need to list only *two* consumer products.   [1 point]

75. The bond between the carbon atoms in ethene is a *double* bond. Hydrocarbons that contain double or triple bonds are said to be unsaturated. Refer to the diagram of ethene below:

           [1 point]

76. Oxidation is the loss of electrons. According to the equation given under the diagram, Zn atoms form Zn^{2+} ions, so the oxidation half-reaction is $Zn \rightarrow Zn^{2+} + 2e^-$. Therefore, oxidation occurs in **half-cell 2**.   [1 point]

77. Reduction occurs in half-cell 1. The balanced half-reaction is:

$$Pb^{2+} + 2e^- \rightarrow Pb$$ [1 point]

78. When the switch is closed, the electrons lost by the Zn electrode travel through the wire and enter the Pb electrode. Therefore, the electron flow is **from the Zn electrode to the Pb electrode**. [1 point]

79. According to Reference Table F, compounds containing the OH^- ion are generally insoluble in water. Since Mg^{2+} is not one of the exceptions, **$Mg(OH)_2$ is insoluble in water**. [1 point]

80. In each case, divide the volume of HCl by the mass of the antacid tablet. Note that only the setup is required:

$$X: \frac{25.20 \text{ mL}}{2.00 \text{ g}}$$

$$Y: \frac{18.65 \text{ mL}}{1.20 \text{ g}}$$

$$Z: \frac{22.50 \text{ mL}}{1.75 \text{ g}}$$ [1 point]

81. Solve each of the fractions given in Question 80. **Antacid Y** (15.5 mL/g) is most effective at neutralizing HCl. [1 point]

82. Risks associated with the use of radium include bone tumors, bone marrow damage, leukemia, and anemia.

Note that only one risk needs to be stated to receive credit. [1 point]

83. According to Reference Table N, radium-226 is an alpha emitter. The symbol for an alpha particle can be obtained from Reference Table O. In order to balance the nuclear equation, the sum of the atomic numbers and the sum of the mass numbers on both sides of the arrow must be equal:

$$^{226}_{88}\text{Ra} \rightarrow {}^{222}_{86}\text{Rn} + {}^{4}_{2}\text{He}$$ [1 point]

84. Refer to the Periodic Table of the Elements. Both radium and calcium are located in Group 2. Elements that are located in the same periodic group have similar chemical properties. [1 point]

85. According to Reference Table N, the half-life of radium-226 is 1600 years. In one half-life period, one-half of the original radioisotope will remain unchanged. Therefore, in order for 1.0 gram of radium-226 to be reduced to 0.50 gram, **1600 years** must pass. [1 point]

Mark (✓) the questions you answered correctly. Count the number of checks and follow the formulas given to determine your score on each topic.

Core Area	☐ Questions Answered Correctly

50, 58, 59, 69, 81

Section M—Math Skills
Number of checks ÷ 5 × 100 = _____%

1, 2, 3, 33, 34, 35, 36

Section I—Atomic Concepts
Number of checks ÷ 7 × 100 = _____%

4, 6, 32, 51, 84

Section II—Periodic Table
Number of checks ÷ 5 × 100 = _____%

8, 37, 52, 61, 62, 66

Section III—Moles/Stoichiometry
Number of checks ÷ 6 × 100 = _____%

5, 7, 9, 10, 38, 41, 60, 70, 71, 72

Section IV—Chemical Bonding
Number of checks ÷ 10 × 100 = _____%

11, 12, 13, 14, 15, 16, 17, 39, 40, 42, 43, 53, 54, 55, 56, 57, 62, 79

Section V—Physical Behavior of Matter
Number of checks ÷ 18 × 100 = _____%

18, 19, 20, 21, 45, 63, 67

Section VI—Kinetics and Equilibrium
Number of checks ÷ 7 × 100 = _____%

22, 23, 44, 46, 65, 73, 74, 75

Section VII—Organic Chemistry
Number of checks ÷ 8 × 100 = _____%

24, 25, 26, 47, 68, 76, 77, 78

Section VIII—Oxidation-Reduction
Number of checks ÷ 8 × 100 = _____%

27, 28, 29, 30, 48, 80

Section IX—Acids, Bases, and Salts
Number of checks ÷ 6 × 100 = _____%

31, 49, 82, 83, 85

Section X—Nuclear Chemistry
Number of checks ÷ 5 × 100 = _____%

Examination
June 2005

Chemistry
The Physical Setting

PART A

Answer all questions in this part.

Directions (1–33): For *each* statement or question, write in the answer space the *number* of the word or expression that, of those given, best completes the statement or answers the question. Some questions may require the use of the *Reference Tables for Physical Setting/Chemistry*.

1 In the modern wave-mechanical model of the atom, the orbitals are regions of the most probable location of

(1) protons   (3) electrons
(2) neutrons   (4) positrons   1 __3__

2 Compared to a proton, an electron has

(1) a greater quantity of charge and the same sign
(2) a greater quantity of charge and the opposite sign
(3) the same quantity of charge and the same sign
(4) the same quantity of charge and the opposite sign   2 __4__

3 Which two notations represent atoms that are isotopes of the same element?

(1) $^{121}_{50}$Sn and $^{119}_{50}$Sn   (3) $^{19}_{8}$O and $^{19}_{9}$F

(2) $^{121}_{50}$Sn and $^{121}_{50}$Sn   (4) $^{39}_{17}$Cl and $^{39}_{19}$K   3 __1__

4 The elements in Period 5 on the Periodic Table are arranged from left to right in order of

(1) decreasing atomic mass
(2) decreasing atomic number
(3) increasing atomic mass
(4) increasing atomic number   4 __4__

5 Which list of elements contains a metal, a metalloid, and a nonmetal?

(1) Zn, Ga, Ge   (3) Cd, Sb, I
(2) Si, Ge, Sn   (4) F, Cl, Br   5 __3__

6 An example of a physical property of an element is the element's ability to

(1) react with an acid
(2) react with oxygen
(3) form a compound with chlorine
(4) form an aqueous solution   6 __4__

7 Which element is malleable and conducts electricity?

(1) iron   (3) sulfur
(2) iodine   (4) phosphorus   7 __1__

8 At STP, solid carbon can exist as graphite or as diamond. These two forms of carbon have

(1) the same properties and the same crystal structures
(2) the same properties and different crystal structures
(3) different properties and the same crystal structures
(4) different properties and different crystal structures

8 __4__

9 What is the formula of titanium(II) oxide?

$Ti^{+2} O^{-2}$

(1) TiO   (3) Ti_2O
(2) TiO_2 (4) Ti_2O_3

9 __1__

10 Which substance can be decomposed by a chemical change?

(1) calcium   (3) copper
(2) potassium   (4) ammonia

10 __4__

11 As a chlorine atom becomes a negative ion, the atom

(1) gains an electron and its radius increases
(2) gains an electron and its radius decreases
(3) loses an electron and its radius increases
(4) loses an electron and its radius decreases

11 __1__

12 Based on Reference Table S, the atoms of which of these elements have the strongest attraction for electrons in a chemical bond?

(1) N   (3) P
(2) Na   (4) Pt

12 __1__

13 Which terms are used to identify pure substances?

   (1) an element and a mixture
   (2) an element and a compound
   (3) a solution and a mixture
   (4) a solution and a compound   13 _2_

14 The solubility of $KClO_3(s)$ in water increases as the

   (1) temperature of the solution increases
   (2) temperature of the solution decreases
   (3) pressure on the solution increases
   (4) pressure on the solution decreases   14 _1_

15 Compared to a 0.1 M aqueous solution of NaCl, a 0.8 M aqueous solution of NaCl has a

   (1) higher boiling point and a higher freezing point
   (2) higher boiling point and a lower freezing point
   (3) lower boiling point and a higher freezing point
   (4) lower boiling point and a lower freezing point   15 _2_

16 The kinetic molecular theory assumes that the particles of an ideal gas

   (1) are in random, constant, straight-line motion
   (2) are arranged in a regular geometric pattern
   (3) have strong attractive forces between them
   (4) have collisions that result in the system losing energy   16 _1_

17 In which process does a solid change directly into a vapor?

   (1) condensation   (3) deposition
   (2) sublimation   (4) solidification   17 _2_

18 Which statement must be true about a chemical system at equilibrium?

(1) The forward and reverse reactions stop.

(2) The concentration of reactants and products are equal.

(3) The rate of the forward reaction is equal to the rate of the reverse reaction.

(4) The number of moles of reactants is equal to the number of moles of product.

18 ___3___

19 Adding a catalyst to a chemical reaction results in

(1) a decrease in activation energy and a decrease in the reaction rate

(2) a decrease in activation energy and an increase in the reaction rate

(3) an increase in activation energy and a decrease in the reaction rate

(4) an increase in activation energy and an increase in the reaction rate

19 ___2___

20 Systems in nature tend to undergo changes toward

(1) lower energy and lower entropy

(2) lower energy and higher entropy

(3) higher energy and lower entropy

(4) higher energy and higher entropy

20 ___2___

21 Which element has atoms that can bond with each other to form long chains or rings?

(1) carbon   (3) oxygen

(2) nitrogen   (4) fluorine

21 ___1___

22 Which formula represents an unsaturated hydro-carbon?

(1) C_2H_6

(3) C_5H_8

(2) C_3H_8

(4) C_6H_{14}

22 __3__

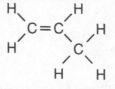

23 Given the structural formula:

What is the IUPAC name of this compound?

(1) propane

(3) propanone

(2) propene

(4) propanal

23 __2__

24 What is the oxidation state of nitrogen in $NaNO_2$?

(1) +1

(3) +3

(2) +2

(4) +4

24 __3 0__

25 The three isomers of pentane have different

(1) formula masses

(2) molecular formulas

(3) empirical formulas

(4) structural formulas

25 __4__

26 Where does oxidation occur in an electrochemical cell?

  (1) at the cathode in both an electrolytic cell and a voltaic cell
  (2) at the cathode in an electrolytic cell and at the anode in a voltaic cell
  (3) at the anode in both an electrolytic cell and a voltaic cell
  (4) at the anode in an electrolytic cell and at the cathode in a voltaic cell   26 __3__

27 Which formula represents an electrolyte?

  (1) CH_3OCH_3 (3) CH_3COOH
  (2) CH_3OH (4) C_2H_5CHO 27 __3__

28 When an Arrhenius acid dissolves in water, the only positive ion in the solution is

  (1) H^+ (3) Na^+
  (2) Li^+ (4) K^+ 28 __1__

29 What is the half-life and decay mode of Rn-222?

  (1) 1.91 days and alpha decay
  (2) 1.91 days and beta decay
  (3) 3.82 days and alpha decay
  (4) 3.82 days and beta decay   29 __3__

30 Which equation represents a transmutation reaction?

  (1) $^{239}_{92}U \rightarrow ^{239}_{92}U + ^{0}_{0}\gamma$

  (2) $^{14}_{6}C \rightarrow ^{14}_{7}N + ^{0}_{-1}e$

  (3) $C_3H_8 + 5O_2 \rightarrow 3CO_2 + 4H_2O$

  (4) $nC_2H_4 + \xrightarrow{\text{catalyst}} (-C_2H_4-)_n$ 30 __2__

31 Which equation represents positron decay?

(1) $^{87}_{37}\text{Rb} \rightarrow {}^{0}_{-1}\text{e} + {}^{87}_{38}\text{Sr}$

(2) $^{227}_{92}\text{U} \rightarrow {}^{223}_{90}\text{Th} + {}^{4}_{2}\text{He}$

(3) $^{27}_{13}\text{Al} + {}^{4}_{2}\text{He} \rightarrow {}^{30}_{15}\text{P} + {}^{1}_{0}\text{n}$

(4) $^{11}_{6}\text{C} \rightarrow {}^{0}_{+1}\text{e} + {}^{11}_{5}\text{B}$

31 __4__

32 Which equation represents a fusion reaction?

(1) $\text{H}_2\text{O(g)} \rightarrow \text{H}_2\text{O}(\ell)$

(2) $\text{C(s)} + \text{O}_2\text{(g)} \rightarrow \text{CO}_2\text{(g)}$

(3) $^{2}_{1}\text{H} + {}^{3}_{1}\text{H} \rightarrow {}^{4}_{2}\text{He} + {}^{1}_{0}\text{n}$

(4) $^{235}_{92}\text{U} + {}^{1}_{0}\text{n} \rightarrow {}^{142}_{56}\text{Ba} + {}^{91}_{36}\text{Kr} + 3{}^{1}_{0}\text{n}$

32 __3__

Note that question 33 has only three choices.

33 An electron in an atom moves from the ground state to an excited state when the energy of the electron

(1) decreases
(2) increases
(3) remains the same

33 __2__

PART B–1

Answer all questions in this part.

Directions (34–50): For *each* statement or question, write in the answer space the *number* of the word or expression that, of those given, best completes the statement or answers the question. Some questions may require the use of the *Reference Tables for Physical Setting/Chemistry*.

34 Which symbol represents a particle that has the same total number of electrons as S^{2-}?

(1) O^{2-} (3) Se^{2-}

(2) Si   (4) Ar   34_____

35 The data table below shows elements Xx, Yy, and Zz from the same group on the Periodic Table.

Element	**Atomic Mass** (atomic mass unit)	**Atomic Radius** (pm)
Xx	69.7	141
Yy	114.8	?
Zz	204.4	171

What is the most likely atomic radius of element Yy?

(1) 103 pm   (3) 166 pm

(2) 127 pm   (4) 185 pm   35_____

36 Which substance has a chemical formula with the same ratio of metal ions to nonmetal ions as in potassium sulfide?

(1) sodium oxide

(2) sodium chloride

(3) magnesium oxide

(4) magnesium chloride   36_____

37 The molecular formula of glucose is $C_6H_{12}O_6$. What is the empirical formula of glucose?

(1) CHO
(2) CH_2O
(3) $C_6H_{12}O_6$
(4) $C_{12}H_{24}O_{12}$

37_____

38 According to Reference Table *F*, which of these compounds is the *least* soluble in water?

(1) K_2CO_3
(2) $KC_2H_3O_2$
(3) $Ca_3(PO_4)_2$
(4) $Ca(NO_3)_2$

38_____

39 A sample of a substance containing only magnesium and chlorine was tested in the laboratory and was found to be composed of 74.5% chlorine by mass. If the total mass of the sample was 190.2 grams, what was the mass of the magnesium?

(1) 24.3 g
(2) 48.5 g
(3) 70.9 g
(4) 142 g

39_____

40 Which molecule contains a nonpolar covalent bond?

$$O=C=O$$
(1)

$$Br-Br$$
(3)

$$C\equiv O$$
(2)

$$Cl-\underset{\underset{Cl}{|}}{\overset{\overset{Cl}{|}}{C}}-Cl$$
(4)

40_____

41 According to Reference Table G, which substance forms an unsaturated solution when 80 grams of the substance is dissolved in 100 grams of H_2O at 10°C?

(1) KI (3) $NaNO_3$

(2) KNO_3 (4) NaCl 41_____

42 What is the concentration of a solution, in parts per million, if 0.02 gram of Na_3PO_4 is dissolved in 1000 grams of water?

(1) 20 ppm (3) 0.2 ppm

(2) 2 ppm (4) 0.02 ppm 42_____

43 Given the simple representations for atoms of two elements:

$\bigcirc$ = an atom of an element

$\bullet$ = an atom of a different element

Which particle diagram represents molecules of only one compound in the gaseous phase?

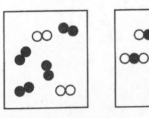

(1) (3)

(2) (4) 43_____

44 Given the balanced equation:

$$KNO_3(s) + 34.89 \text{ kJ} \xrightarrow{H_2O} K^+(aq) + NO_3^-(aq)$$

Which statement best describes this process?

(1) It is endothermic and entropy increases.
(2) It is endothermic and entropy decreases.
(3) It is exothermic and entropy increases.
(4) It is exothermic and entropy decreases.

44_____

45 A 1.0-gram piece of zinc reacts with 5 milliliters of HCl(aq). Which of these conditions of concentration and temperature would produce the greatest rate of reaction?

(1) 1.0 M HCl(aq) at 20.°C
(2) 1.0 M HCl(aq) at 40.°C
(3) 2.0 M HCl(aq) at 20.°C
(4) 2.0 M HCl(aq) at 40.°C

45_____

46 At STP, fluorine is a gas and iodine is a solid. This observation can be explained by the fact that fluorine has

(1) weaker intermolecular forces of attraction than iodine
(2) stronger intermolecular forces of attraction than iodine
(3) lower average kinetic energy than iodine
(4) higher average kinetic energy than iodine

46_____

47 Given the structural formula:

$$H-\overset{\overset{\displaystyle H}{|}}{\underset{\underset{\displaystyle H}{|}}{C}}-\overset{\overset{\displaystyle H}{|}}{\underset{\underset{\displaystyle H}{|}}{C}}-O-\overset{\overset{\displaystyle H}{|}}{\underset{\underset{\displaystyle H}{|}}{C}}-\overset{\overset{\displaystyle H}{|}}{\underset{\underset{\displaystyle H}{|}}{C}}-H$$

The compound represented by this formula can be classified as an

(1) organic acid   (3) ester
(2) ether   (4) aldehyde   47____

48 Sulfuric acid, H_2SO_4(aq), can be used to neutralize barium hydroxide, $Ba(OH)_2$(aq). What is the formula for the salt produced by this neutralization?

(1) BaS   (3) $BaSO_3$
(2) $BaSO_2$ (4) $BaSO_4$ 48____

49 Given the balanced ionic equation:

$$Zn(s) + Cu^{2+}(aq) \rightarrow Zn^{2+}(aq) + Cu(s)$$

Which equation represents the oxidation half-reaction?

(1) $Zn(s) + 2e^- \rightarrow Zn^{2+}(aq)$
(2) $Zn(s) \rightarrow Zn^{2+}(aq) + 2e^-$
(3) $Cu^{2+}(aq) \rightarrow Cu(s) + 2e^-$
(4) $Cu^{2+}(aq) + 2e^- \rightarrow Cu(s)$ 49____

50 In which solution will thymol blue indicator appear blue?

(1) 0.1 M CH_3COOH (3) 0.1 M HCl
(2) 0.1 M KOH   (4) 0.1 M H_2SO_4 50____

PART B-2

Answer all questions in this part.

Directions (51–64): Record your answers in the spaces provided on the answer sheet provided in the back. Some questions may require the use of the *Reference Tables for Physical Setting/Chemistry*.

Base your answers to questions 51 and 52 on the diagram below, which represents an atom of magnesium-26 in the ground state.

Mg-26 nucleus

51 What is the total number of valence electrons in an atom of Mg-26 in the ground state? [1]

52 On the diagram *on your answer sheet*, write an appropriate number of electrons in *each* shell to represent a Mg-26 atom in an excited state. Your answer may include additional shells. [1]

_____

53 Explain, in terms of atomic structure, why germanium is chemically similar to silicon. [1]

54 Given the balanced equation:

$$4Al(s) + 3O_2(g) \rightarrow 2Al_2O_3(s)$$

What is the total number of moles of $O_2(g)$ that must react completely with 8.0 moles of $Al(s)$ in order to form $Al_2O_3(s)$? [1]

Base your answers to questions 55 and 56 on the balanced equation below.

$$2Na(s) + Cl_2(g) \rightarrow 2NaCl(s)$$

55 In the box *on your answer sheet*, draw a Lewis electron-dot diagram for a molecule of chlorine, Cl_2. [1]

56 Explain, in terms of electrons, why the bonding in NaCl is ionic. [1]

Base your answers to questions 57 and 58 on the information below:

Given the reaction at equilibrium:

$$2NO_2(g) + 7H_2(g) \rightleftharpoons 2NH_3(g) + 4H_2O(g) + 1127 \text{ kJ}$$

57 On the diagram *on your answer sheet*, complete the potential energy diagram for the forward reaction. Be sure your drawing shows the activation energy and the potential energy of the products. [2]

58 Explain, in terms of Le Chatelier's principle, why the concentration of $NH_3(g)$ *decreases* when the temperature of the equilibrium system increases. [1]

Base your answers to questions 59 and 60 on the information below.

Given the reaction between 1-butene and chlorine gas:

$$C_4H_8 + Cl_2 \rightarrow C_4H_8Cl_2$$

59 Which type of chemical reaction is represented by this equation?   [1]

60 In the space *on your answer sheet*, draw the structural formula of the product 1,2-dichlorobutane. [1]

Base your answers to questions 61 through 64 on the information below, which relates the numbers of neutrons and protons for specific nuclides of C, N, Ne, and S.

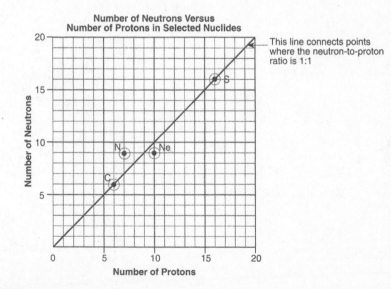

Number of Neutrons Versus Number of Protons in Selected Nuclides

This line connects points where the neutron-to-proton ratio is 1:1

61 Using the point plotted on the graph for neon, complete the table *on your answer sheet.* [1]

62 Explain, in terms of atomic particles, why S-32 is a stable nuclide. [1]

63 Using the point plotted on the graph for nitrogen, what is the neutron-to-proton ratio of this nuclide? [1]

64 Based on Reference Table N, complete the decay equation for N-16 *on your answer sheet.* [1]

PART C

Answer all questions in this part.

Directions (65–83): Record your answers on the answer sheet provided in the back. Some questions may require the use of the *Reference Tables for Physical Setting/Chemistry.*

65 In the early 1900s, experiments were conducted to determine the structure of the atom. One of these experiments involved bombarding gold foil with alpha particles. Most alpha particles passed directly through the foil. Some, however, were deflected at various angles. Based on this alpha particle experiment, state *two* conclusions that were made concerning the structure of an atom.   [2]

Base your answers to questions 66 through 70 on the information below.

A substance is a solid at 15°C. A student heated a sample of the solid substance and recorded the temperature at one-minute intervals in the data table below.

Time (min)	0	1	2	3	4	5	6	7	8	9	10	11	12
Temperature (°C)	15	32	46	53	53	53	53	53	53	53	53	60	65

66 On the grid *on your answer sheet*, mark an appropriate scale on the axis labeled "Temperature (°C)." An appropriate scale is one that allows a trend to be seen.   [1]

67 Plot the data from the data table. Circle and connect the points. [1]

Example:

68 Based on the data table, what is the melting point of this substance? [1]

69 What is the evidence that the average kinetic energy of the particles of this substance is increasing during the first three minutes? [1]

70 The heat of fusion for this substance is 122 joules per gram. How many joules of heat are needed to melt 7.50 grams of this substance at its melting point? [1]

Base your answers to questions 71 through 73 on the diagram of a voltaic cell and the balanced ionic equation below.

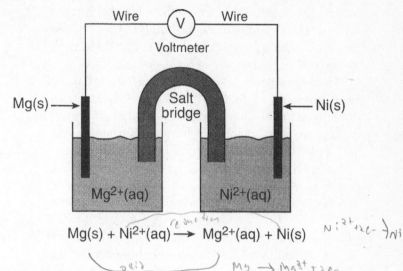

$$Mg(s) + Ni^{2+}(aq) \longrightarrow Mg^{2+}(aq) + Ni(s)$$

(handwritten annotations: reduction, oxid, $Ni^{2+} + 2e^- \rightarrow Ni$, $Mg \rightarrow Mg^{2+} + 2e^-$)

71 What is the total number of moles of electrons needed to completely reduce 6.0 moles of $Ni^{2+}(aq)$ ions? [1]

72 Identify *one* metal from Reference Table J that is more easily oxidized than Mg(s). [1]

73 Explain the function of the salt bridge in the voltaic cell. [1]

Base your answers to questions 74 through 76 on the passage below.

Acid rain is a problem in industrialized countries around the world. Oxides of sulfur and nitrogen are formed when various fuels are burned. These oxides dissolve in atmospheric water droplets that fall to earth as acid rain or acid snow.

While normal rain has a pH between 5.0 and 6.0 due to the presence of dissolved carbon dioxide, acid rain often has a pH of 4.0 or lower. This level of acidity can damage trees and plants, leach minerals from the soil, and cause the death of aquatic animals and plants.

If the pH of the soil is too low, then quicklime, CaO, can be added to the soil to increase the pH. Quicklime produces calcium hydroxide when it dissolves in water.

74 Balance the neutralization equation *on your answer sheet*, using the smallest whole-number coefficients. [1]

75 A sample of wet soil has a pH of 4.0. After the addition of quicklime, the H^+ ion concentration of the soil is $\frac{1}{100}$ of the original H^+ ion concentration of the soil. What is the new pH of the soil sample? [1]

76 Samples of acid rain are brought to a laboratory for analysis. Several titrations are performed and it is determined that a 20.0-milliliter sample of acid rain is neutralized with 6.50 milliliters of 0.010 M NaOH. What is the molarity of the H^+ ions in the acid rain? [1] $M_A V_A = M_B V_B$

$① \quad 20 \times 6.5 = \dfrac{10 \times}{10}$

$10°$

Base your answers to questions 77 through 79 on the information and diagrams below.

Cylinder *A* contains 22.0 grams of $CO_2(g)$ and cylinder *B* contains $N_2(g)$. The volumes, pressures, and temperatures of the two gases are indicated under each cylinder.

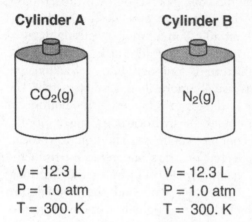

Cylinder A   **Cylinder B**

$CO_2(g)$ $N_2(g)$

V = 12.3 L   V = 12.3 L
P = 1.0 atm   P = 1.0 atm
T = 300. K   T = 300. K

77 What is the total number of moles of $CO_2(g)$ in cylinder *A*? [1]

78 Explain why the number of molecules of $N_2(g)$ in cylinder *B* is the same as the number of molecules of $CO_2(g)$ in cylinder *A*. [1]

79 The temperature of the $CO_2(g)$ is increased to 450. K and the volume of cylinder *A* remains constant. In the space *on your answer sheet*, show a correct numerical setup for calculating the new pressure of the $CO_2(g)$ in cylinder *A*. [1]

Base your answers to questions 80 through 83 on the information and diagram below and on your knowledge of chemistry.

Crude oil is a mixture of many hydrocarbons that have different numbers of carbon atoms. The use of a fractionating tower allows the separation of this mixture based on the boiling points of the hydrocarbons.

To begin the separation process, the crude oil is heated to about 400°C in a furnace, causing many of the hydrocarbons of the crude oil to vaporize. The vaporized mixture is pumped into a fractionating tower that is usually more than 30 meters tall. The temperature of the tower is highest at the bottom. As vaporized samples of hydrocarbons travel up the tower, they cool and condense. The liquid hydrocarbons are collected on trays and removed from the tower. The diagram below illustrates the fractional distillation of the crude oil and the temperature ranges in which the different hydrocarbons condense.

Distillation of Crude Oil

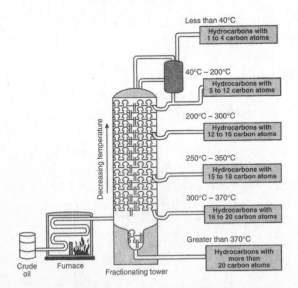

80 State the trend between the boiling point of the hydrocarbons contained in the crude oil and the number of carbon atoms in these molecules.   [1]

81 Describe the relationship between the strength of the intermolecular forces and the number of carbon atoms in the different hydrocarbon molecules.   [1]

82 Write an IUPAC name of *one* saturated hydrocarbon that leaves the fractionating tower at *less than* 40°C. [1]

83 How many hydrogen atoms are present in one molecule of octane?   [1]

Answer Sheet
June 2005

Chemistry
The Physical Setting

PART B–2

51 _____

52

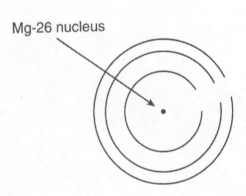

Mg-26 nucleus

53 _____
_____
_____
_____

54 _____ mol

55

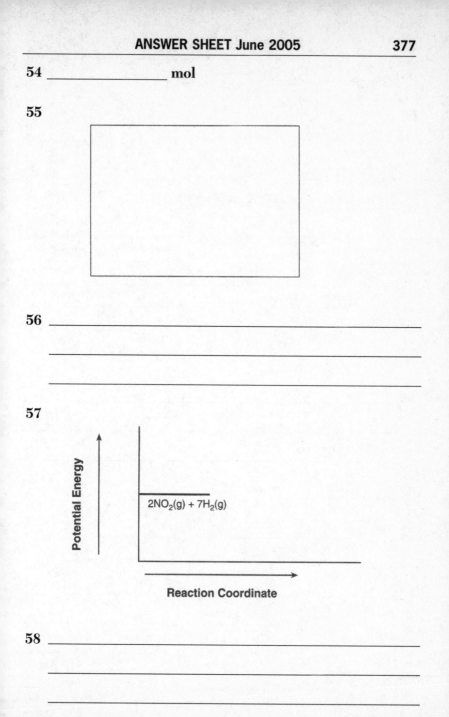

56 _____

_____

_____

57

58 _____

_____

_____

59 _____

60

61

Element	Number of Protons	Number of Neutrons	Mass Number	Nuclide
C	6	6	12	C-12
N	7	9	16	N-16
Ne	10			
S	16	16	32	S-32

62 _____

_____

_____

63 _____

64 $^{16}_{7}N \rightarrow$ _____ + _____

PART C

65 Conclusion 1: _____

_____

_____

Conclusion 2: _____

_____

_____

66 and **67**

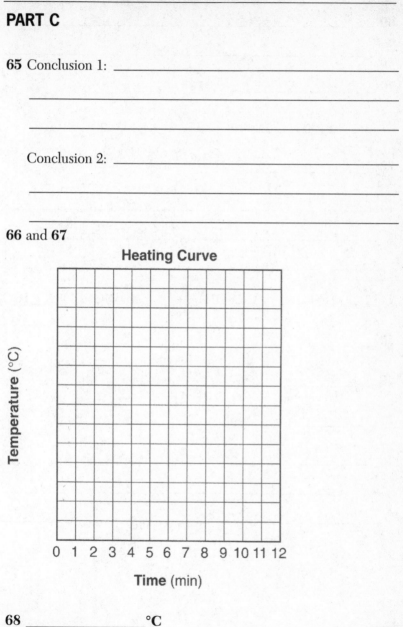

Heating Curve

Temperature (°C)

0 1 2 3 4 5 6 7 8 9 10 11 12

Time (min)

68 _____ **°C**

69 _____

_____

70 _____ **J**

71 _____ **mol**

72 _____

73 _____

_____

_____

74 _____ HNO_3+ _____ $Ca(OH)_2 \rightarrow$ _____ $Ca(NO_3)_2$+ _____ H_2O

75 _____

76 _____ **M**

77 _____ **mol**

78 _____

_____

_____

79

80 _____

_____

81 _____

_____

82 _____

83 _____

Answers
June 2005

Chemistry
The Physical Setting

Answer Key

PART A

1. 3	**7.** 1	**13.** 2	**19.** 2	**25.** 4	**31.** 4
2. 4	**8.** 4	**14.** 1	**20.** 2	**26.** 3	**32.** 3
3. 1	**9.** 1	**15.** 2	**21.** 1	**27.** 3	**33.** 2
4. 4	**10.** 4	**16.** 1	**22.** 3	**28.** 1	
5. 3	**11.** 1	**17.** 2	**23.** 2	**29.** 3	
6. 4	**12.** 1	**18.** 3	**24.** 3	**30.** 2	

PART B–1

34. 4	**38.** 3	**42.** 1	**46.** 1	**50.** 2
35. 3	**39.** 2	**43.** 3	**47.** 2	
36. 1	**40.** 3	**44.** 1	**48.** 4	
37. 2	**41.** 1	**45.** 4	**49.** 2	

Answers Explained

PART A

1. **3** The modern wave-mechanical model of the atom defines an orbital as a region in which a given electron is most likely to be located.

2. **4** An electron is the basic particle that carries a negative charge. A proton is the basic particle that carries a positive charge. The quantity of charge on a proton is equal to the quantity of charge on an electron.

3. **1** Isotopes of an element contain the same number of protons but differing numbers of neutrons. Therefore, two isotopes of the same element will have the same atomic number (in this case, 50) and different mass numbers (in this case, 121 and 119).

Wrong Choices Explained:
(2) This pair represents the same atom.
(3), (4) These pairs represent different elements having the same mass numbers.

4. **4** Refer to the Periodic Table of the Elements. Elements are placed in order of increasing atomic number.

Wrong Choice Explained:
(3) Element 53 (I) has a smaller atomic mass than element 52 (Te).

5. **3** Refer to the Periodic Table of the Elements. Cd, located in Group 12 and to the left of the metal-nonmetal line, is a metal; Sb, located in Group 15 and on the metal-nonmetal line, is a metalloid; I, located in Group 17 and to the right of the metal-nonmetal line, is a nonmetal.

Wrong Choices Explained:
(1) Zn and Ga are metals; Ge is a metalloid.
(2) Si and Ge are metalloids; Sn is a metal.
(4) F, Cl, and Br are nonmetals.

6. **4** A physical property is one that can be measured without changing the identity of the substance involved. When an element forms an aqueous solu-

tion, the element retains its identity. For example, when Cl_2 dissolves in water, the Cl_2 molecules remain intact.

Wrong Choices Explained:
(1), (2), (3) In each case, the element changes its identity from neutral atoms to ions.

7. **1** Malleability and conductivity are properties found in metallic substances such as iron.

8. **4** Graphite and diamond have distinctly different crystal stuctures. Therefore, they will have different properties as well.

9. **1** The charges of the titanium (II) and oxide ions are, respectively, 2+ and 2–. In a (neutral) compound, the sum of the charges must add to zero. Therefore, the correct formula is TiO.

10. **4** Compounds can be decomposed by chemical change, while elements cannot undergo further change. Of the choices given, only choice (4), ammonia, is a compound.

11. **1** Atoms form ions by gaining or losing electrons. When a chlorine atom becomes a negative ion, it does so by gaining a single electron. The presence of this additional electron causes the radius of the ion to be larger than the radius of its parent atom.

12. **1** Electronegativity is defined as the attraction for electrons in a chemical bond. The higher the electronegativity, the stronger the attraction. According to Reference Table S, choice (1), N, has the highest electronegativity (3.0) of the four choices given.

13. **2** A pure substance is defined as either an element or a compound.

14. **1** See Reference Table G. As the temperature increases, the solubility of $KClO_3$ increases. Pressure has virtually no effect on the solubility of solids in water.

15. **2** The boiling and freezing points of a solution depend on the number of dissolved solute particles in the solution. As the number of dissolved particles increases, the boiling point increases and the freezing point decreases. Since a 0.8 M solution of NaCl has more dissolved particles than a 0.1 M solu-

tion of NaCl, the 0.8 M solution will have a higher boiling point and a lower freezing point than the 0.1 M solution.

16. **1** According to the kinetic molecular theory, an ideal gas consists of particles that are in constant, random, straight-line motion. Moreover, these particles have no attractive or repulsive forces between them, and they always collide elastically (that is, without loss of energy). Of the choices given, only choice (1) meets all of these criteria.

17. **2** Sublimation is defined as the direct change of a solid to a vapor (gas). Solid carbon dioxide (dry ice) is an example of a substance that sublimes at room temperature and pressure.

Wrong Choices Explained:
   (1) Condensation is the change of a vapor into a liquid.
   (3) Deposition is the change of a vapor directly into a solid.
   (4) Solidification is the change of a liquid into a solid.

18. **3** A chemical system is considered to be at equilibrium when the rates of its forward and reverse reactions are equal.

19. **2** A catalyst increases the rate of a reaction by providing an alternative pathway for the reaction to occur. (This is somewhat like certain roads in Europe that pass through mountains rather than go around them.) The alternative pathway has a lower activation energy than the original pathway. As a result, molecules with lower kinetic energies are capable of reacting and the speed of the reaction is increased.

20. **2** Changes in nature are directed by two factors: energy and entropy. Systems in nature will tend to undergo spontaneous changes if the energy of the system is decreased and the entropy (disorder) of the system is increased.

21. **1** Of the choices given, only choice (1), carbon, is unique in that its atoms can bond with each other to form long chains or rings. The reasons for this uniqueness lie in the fact that carbon is located in the middle group of the Periodic Table of the Elements, has four valence electrons, and can alter its valence orbitals in a number of ways.

22. **3** An unsaturated hydrocarbon is one that contains a double or a triple bond (that is, an alkene or an alkyne). See Reference Table Q. Of the choices given, only the formula for choice (3), C_5H_8, matches the general formula

(C_nH_{2n-2}) of an unsaturated hydrocarbon. The formula C_5H_8 represents an alkyne.

Wrong Choices Explained:

(1), (2), (4) Each of these formulas matches the general formula (C_nH_{2n+2}) of an alkane, a *saturated* hydrocarbon.

23. **2** See Reference Tables P and Q. The structural formula has three carbon atoms: it will have the prefix *prop-*. Moreover, the formula contains one double bond: it will have the suffix *-ene*. Therefore, the IUPAC name will be propene.

Wrong Choices Explained:

(1) Propane is the name of an alkane.
(3) Propanone is the name of a ketone; it is not a hydrocarbon.
(4) Propanal is the name of an aldehyde; it is not a hydrocarbon.

24. **3** Since $NaNO_2$ is a neutral compound, the sum of the oxidation numbers must add to zero. Na is located in Group 1 and has an oxidation number of +1. Oxygen is located in Group 16 and has an oxidation number of –2. Since two oxygen atoms are present, the contribution of oxygen is –4. Nitrogen must have an oxidation number of +3 in order to bring the total to zero.

25. **4** Isomers of a compound have the same molecular formula (as well as the same empirical formula and molecular mass) but different structural formulas.

26. **3** The anode is defined to be the electrode at which oxidation occurs in an electrochemical cell. This definition holds for both voltaic and electrolytic cells.

27. **3** An electrolyte is a substance that produces ions in aqueous solution, and as a result, the solution conducts electricity. When ethanoic acid (CH_3COOH) is dissolved in water, a small quantity of H^+ and CH_3COO^- ions are produced.

Wrong Choices Explained:

(1), (2), (4) These compounds are all *nonelectrolytes*. They produce no ions when dissolved in water, and the resulting solutions do not conduct electricity.

28. **1** An Arrhenius acid is a substance that dissolves in water to produce H^+ ions as the only positive ions in solution.

29. **3** See Reference Tables N and O. Rn-222 has a half-life of 3.82 days and undergoes alpha (α) decay.

30. **2** In a transmutation reaction, a type of nuclear reaction, one element is converted into another. Of the choices given, only the equation in choice (2), the beta decay of carbon-14, is a transmutation reaction.

Wrong Choices Explained:
(1) This equation does not involve the change of an element.
(3) This equation represents *combustion,* which is not a nuclear reaction.
(4) This equation represents *polymerization*, which is not a nuclear reaction.

31. **4** See Reference Table O. A positron may be represented by the symbol $_{+1}^{0}e$. Of the choices given, only choice (4), $_{6}^{11}C \rightarrow \,_{+1}^{0}e + \,_{5}^{11}B$, illustrates the emission of a positron.

32. **3** In a fusion reaction, lighter nuclei join to form a heavier nucleus. In choice (3), the lighter nuclei $_{1}^{2}H$ and $_{1}^{3}H$ join to form the heavier nucleus $_{2}^{4}He$.

Wrong Choices Explained:
(1), (2) These equations do not represent nuclear reactions.
(4) This equation represents nuclear *fission*, in which a heavier nucleus is broken into lighter nuclei.

33. **2** The ground state is the lowest energy state that an atom can have. When an electron moves into a higher energy level, the atom is said to be excited and the electron's energy increases.

PART B–1

34. **4** An atom of S has 16 electrons, so the S^{2-} ion has 18 electrons. Of the choices listed, only choice (4), Ar, has 18 electrons.

Wrong Choices Explained:
(1) O^{2-} has 10 electrons.
(2) Si has 14 electrons.
(3) Se^{2-} has 36 electrons.

35. **3** Refer to the Periodic Table of the Elements. Elements Xx, Yy, and Zz are all located in Group 13 and are, respectively, Ga, In, and Tl. The unknown atomic radius belongs to the element In, (atomic number 49). Use Reference Table S. The atomic radius of In is 166 pm.

36. **1** The formula for potassium sulfide is K_2S, in which the ratio of metal ions to nonmetal ions is 2:1. Choice (1), sodium oxide, has the formula Na_2O and the same metal ion–nonmetal ion ratio.

Wrong Choices Explained:
(2) In sodium chloride (NaCl), the metal ion–nonmetal ion ratio is 1:1.
(3) In magnesium oxide (MgO), the metal ion–nonmetal ion ratio is 1:1.
(4) In magnesium chloride ($MgCl_2$), the metal ion–nonmetal ion ratio is 1:2.

37. **2** An empirical formula is one whose atoms are expressed in *smallest* whole-number ratios. To obtain the empirical formula of $C_6H_{12}O_6$, divide the formula by 6: CH_2O.

38. **3** According to Reference Table F, compounds containing the phosphate (PO_4^{3-}) ion are generally insoluble (the exceptions being phosphates combined with NH_4^+ or Group 1 ions). Therefore, $Ca_3(PO_4)_2$ is insoluble in water.

Wrong Choices Explained:
(1), (2) All compounds containing Group 1 ions (such as K^+) or the acetate ion ($C_2H_3O_2^-$) are soluble in water.
(4) All compounds containing the nitrate ion (NO_3^-) are soluble in water.

39. **2** If the compound contains 74.5% chlorine by mass, it must contain 25.5% magnesium by mass (100.0% − 74.5%). The mass of magnesium present is found by multiplying the total mass (190.2 g) by the fraction of magnesium (25.5% or 0.255) in the compound:

$$190.2 \text{ g} \times 0.255 = \textbf{48.5 g}$$

40. **3** Nonpolar bonds have electronegativity differences of zero. See Reference Table S. Of the choices given, only choice (3), Br–Br, has an electronegativity difference of zero.

Wrong Choices Explained:

(1), (2) The electronegativity difference between C and O is 0.8 (a polar covalent bond).

(4) The electronegativity difference between C and Cl is 0.6 (a polar covalent bond).

41. **1** According to Reference Table G, a solution containing 100 grams of H_2O will be unsaturated if the quantity of solute lies *below* the solubility curve for that substance. In choice (1), at 10°C, 80 grams of KI lies well below the solubility curve for KI, so the solution will be unsaturated.

Wrong Choices Explained:

(2), (4) At 10°C, 80 grams of KNO_3 and NaCl lie *above* their respective solubility curves. These solutions will be saturated and contain an excess of undissolved solute.

(3) At 10°C, 80 grams of $NaNO_3$ lies *on* the solubility curve. This solution will be exactly saturated.

42. **1** See the concentration formulas on Reference Table T. In this problem, the solution is so dilute that the mass of the water can be used as the mass of the solution.

$$\text{parts per million} = \frac{\text{grams of solute}}{\text{grams of solution}} \times 1\,000\,000$$

$$= \frac{0.02 \text{ g}}{1000 \text{ g}} \times 1\,000\,000 = \textbf{20 ppm}$$

43. **3** In each case, the molecules are separated and fill their respective containers. Therefore, they are all in the gas phase. Use the following letters to represent the elements:

○ = Element A

● = Element B

The diagram corresponding to choice (3) contains molecules of only one compound: A_2B.

Wrong Choices Explained:
(1) This diagram contains molecules of element A as A_3.
(2) This diagram contains elements A and B as A_2 and B_2.
(4) This diagram contains two compounds: A_2B and AB.

44. **1** Since 34.89 kJ appears on the *left side* of the equation, it indicates that heat has been absorbed by the system. This is an endothermic reaction. In the dissolving process, the KNO_3 crystal is separated into ions, which are more randomly arranged in the solution than in the crystal. Therefore, the entropy of the system increases.

45. **4** Reaction rates depend on temperature and the concentration of reactants. A reaction that occurs at 40°C will be faster than the reaction at 20°C. Furthermore, 2.0 M HCl is more concentrated than 1.0 M HCl. Using 2.0 M HCl will produce a faster reaction than using 1.0 M HCl.

46. **1** The phase of a molecular substance at STP depends on the strength of the intermolecular forces of attraction. Molecules with weaker intermolecular forces tend to be gases (such as fluorine or chlorine), while substances with stronger intermolecular forces tend to be liquids or solids (such as bromine or iodine).

47. **2** See Reference Table R. The formula corresponds to the general type R–O–R', an *ether*. The name of this compound is diethyl ether. The general formulas for the other choices (organic acid, ester, and aldehyde) can also be found on Reference Table R.

48. **4** The neutralization reaction is shown below:

$$H_2SO_4(aq) + Ba(OH)_2(aq) \rightarrow 2H_2O(\ell) + \textbf{BaSO}_4(s)$$

49. **2** Oxidation involves the loss of electrons. As a result, the oxidation number of the substance *increases*. Of the choices given, only the half-reaction shown in choice (2) meets both of these criteria.

Wrong Choices Explained:

(1) This equation is an incorrectly written reduction half-reaction.

(3) This equation is an incorrectly written oxidation half-reaction.

(4) This equation is correctly written, but it represents reduction (the gain of electrons).

50. **2** See Reference Table *M*. According to the table, thymol blue will be blue at pH values of 9.6 and above, that is, in a basic solution. Of the choices given, only choice (2), 0.1 M KOH, is a basic solution.

Wrong Choices Explained:

(1), (3), (4) Each of these solutions is *acidic*: the pH will be less than 7.0 and the color of thymol blue will be yellow.

PART B–2

[Point values are indicated in brackets.]

51. Valence electrons are those electrons that occupy the outermost level of an atom. Since the electron configuration for a ground-state magnesium atom is 2-8-2, the atom has a total of 2 valence electrons. [1 point]

52. When an atom becomes excited, one or more electrons are promoted to higher energy levels. Refer to the diagrams below, which represent two such possibilities.

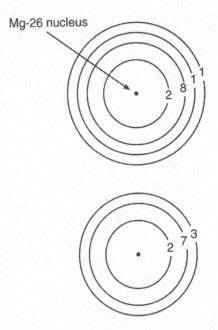

Mg-26 nucleus

[1 point]

53. The chemical similarities between germanium and silicon lie in the fact that both elements have the same number of valence electrons (4). As a result, both elements are located in Group 14 of the Periodic Table of the Elements. [1 point]

54. The coefficients of any balanced equation indicate the relative numbers of moles of reactants and products:

$$8.0 \text{ mol Al} \times \left(\frac{3 \text{ mol O}_2}{4 \text{ mol Al}} \right) = \mathbf{6.0 \text{ mol O}_2}$$

[1 point]

55. The Lewis electron-dot diagram for a molecule of Cl_2 shows two chlorine atoms sharing a single pair of electrons. Each chlorine atom in the molecule has a total of eight electrons. Two possible diagrams are shown below:

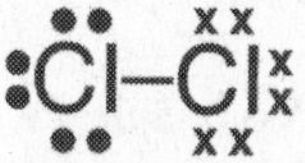

[1 point]

56. NaCl is ionic because the sodium atom transfers its single valence electron to the chlorine atom. As a result, Na^+ and Cl^- ions, each having 8 valence electrons, are formed. The diagram below illustrates the transfer of the valence electron:

[1 point]

57. In this equation, the heat appears on the right side of the equation. It is an *exothermic* reaction, in which the potential energy of the products is lower than the potential energy of the reactants. A labeled diagram is shown below:

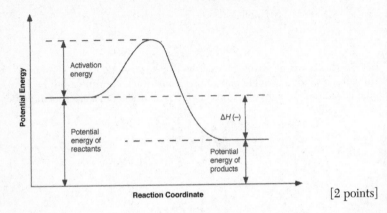

[2 points]

58. According to Le Chatelier's principle, increasing the temperature favors the *endothermic* reaction (that is, the reverse reaction). As the system shifts to the left, the concentration of $NH_3(g)$ will decrease. [1 point]

59. 1-butene contains a double bond. Hydrocarbons containing double and/or triple bonds will undergo an addition reaction when combined with chlorine, bromine, or iodine.

Note that other acceptable responses include chlorination, halogenation, redox, and synthesis, but addition is the preferred answer. [1 point]

60. Two possibilities for the structural formula of 1,2-dichlorobutane are shown below:

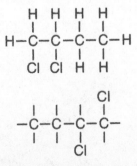

[1 point]

61. The completed table is shown below. Since this isotope of neon has 10 protons and 9 neutrons, its mass number is 19. The nuclide is written as Ne-19 (or as ^{19}Ne).

Element	Number of Protons	Number of Neutrons	Mass Number	Nuclide
C	6	6	12	C-12
N	7	9	16	N-16
Ne	**10**	**9**	**19**	**Ne-19**
S	16	16	32	S-32

Note that the three correct responses are written in boldfaced type. [1 point]

62. The *neutron-to-proton ratio* determines the stability of a nuclide. For lighter elements (including sulfur), a 1:1 ratio is favored. The nuclide S-32 has a 1:1 ratio (16 protons and 16 neutrons). [1 point]

63. According to the point plotted on the graph, this nuclide of nitrogen (N) has 9 neutrons and 7 protons. Therefore, the ratio is 9:7 (or $\frac{9}{7}$). [1 point]

64. See Reference Tables *N* and *O*. The nuclide N-16 undergoes beta decay, which means that an electron is ejected from the nucleus of the nuclide. In order to complete the decay equation, the mass numbers and atomic numbers (or charges) on both sides of the equation must be equal. The complete equation is shown below:

$$^{16}_{7}\text{N} \rightarrow {}^{0}_{-1}\text{e} + {}^{16}_{8}\text{O}$$

[1 point]

PART C

[Point values are indicated in brackets.]

65. Acceptable conclusions include:
- The nucleus is small.
- The nucleus is dense.
- The nucleus is positively charged.
- Most of the atom is empty space.

[2 points]

66. Refer to the graph that appears after question 67. An ideal temperature scale is one that nearly fills the entire graph and allows a trend to be observed. [1 point]

67. One point is awarded for plotting all of the data points correctly (that is, within ±0.3 of a grid space). The completed graph is shown below.

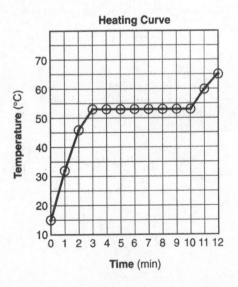

Heating Curve

[1 point]

68. At the melting point, the temperature remains constant (that is, the graph plateaus). For this substance, the melting point is **53°C**. [1 point]

69. Temperature is related to the average kinetic energy of the particles. Since the temperature is increasing during the first three minutes, the average kinetic energy of the particles is also increasing. [1 point]

70. The heat of fusion is the amount of heat needed to melt 1 gram of substance at its melting point. The solution is shown below:

$$7.50 \text{ g} \times \frac{122 \text{ J}}{1 \text{ g}} = \textbf{915 J}$$

[1 point]

71. In order for 1 mole of $Ni^{2+}(aq)$ to be reduced to $Ni(s)$, 2 moles of electrons are needed. The reduction half-reaction is:

$$Ni^{2+}(aq) + 2e^- \rightarrow Ni(s)$$

The solution to the problem is:

$$6 \text{ mol Ni}^{2+} \times \left(\frac{2 \text{ mol e}^-}{1 \text{ mol Ni}^{2+}} \right) = \textbf{12 mol e}^-$$

[1 point]

72. See Reference Table *J*. The metals are listed in descending order. In other words, the higher the metal on the table, the more easily it will be oxidized. Choose any single metal that lies above Mg on this table (Li, Rb, K, Cs, Ba, Sr, Ca, Na). [1 point]

73. The salt bridge completes the electric circuit by permitting ions to flow into the half-cells. As a result, the half-cells remain electrically neutral. [1 point]

74. The balanced neutralization equation (using smallest whole-number coefficients) is:

$$\textbf{2}HNO_3 + Ca(OH)_2 \rightarrow Ca(NO_3)_2 + \textbf{2}H_2O$$

Note that placing the coefficient **1** in front of $Ca(OH)_2$ and $Ca(NO_3)_2$ is acceptable. [1 point]

75. If a sample has a pH of 4.0, by definition its H^+ ion concentration is 1.0×10^{-4} M. If the H^+ ion concentration is reduced to $\frac{1}{100}$ of its initial value, the ion concentration is 1.0×10^{-6} M, a value corresponding to a pH of **6.0**. [1 point]

76. Use the titration equation found on Reference Table T:

$$M_A V_A = M_B V_B$$
$$M_A \cdot (20.0 \text{ mL}) = (0.010 \text{ M}) \cdot (6.50 \text{ mL})$$
$$M_A = \textbf{0.0033 M (or 0.00325 M)}$$

[1 point]

77. One mole of any substance is its molar mass expressed in grams. The molar mass of CO_2 is 44.0 grams. The solution to this problem is shown below:

$$22.0 \text{ g} CO_2 \cdot \left(\frac{1 \text{ mol} CO_2}{44.0 \text{ g} CO_2} \right) = \textbf{0.500 mol } CO_2$$

[1 point]

78. According to Avogadro's principle, equal volumes of gases at the same pressure and temperature contain equal numbers of particles. The pressure, volume, and temperature of both containers of gases are the same. [1 point]

79. Use the combined gas law equation found on Reference Table T:

$$\frac{P_1 V_1}{T_1} = \frac{P_2 V_2}{T_2}$$
$$\frac{(1.0 \text{ atm}) \cdot (12.3 \text{ L})}{(300 \text{ K})} = \frac{P_2 \cdot (12.3 \text{ L})}{(450 \text{ K})}$$
$$P_2 = \frac{450}{300} \textbf{ atm}$$

[1 point]

Note that only the setup is necessary, not the numerical solution.

80. As the number of carbon atoms in these molecules increases, the boiling point increases. [1 point]

81. Boiling points are related to intermolecular forces: as the intermolecular forces among the molecules become stronger, the boiling point increases. Therefore, as the number of carbon atoms increases, the strength of the intermolecular forces increases. [1 point]

82. Any alkane with 4 carbons or less will boil at a temperature less than 40°C. Any one of the following is an acceptable answer: methane, ethane, propane, methylpropane, or butane. [1 point]

83. See Reference Tables P and Q. Octane is an alkane with 8 carbon atoms. Since the general formula for an alkane is C_nH_{2n+2}, the number of hydrogen atoms in octane will be:

$$2 \cdot 8 + 2 = \mathbf{18}$$

[1 point]

Mark (✓) the questions you answered correctly. Count the number of checks and follow the formulas given to determine your score on each topic.

Core Area	☐ Questions Answered Correctly

61, 63, 66, 67

Section M—Math Skills
☐   Number of checks ÷ 4 × 100 =   ____%

1, 2, 3, 33, 51, 52, 65

Section I—Atomic Concepts
☐   Number of checks ÷ 7 × 100 =   ____%

4, 5, 6, 7, 8, 35, 53, 72

Section II—Periodic Table
☐   Number of checks ÷ 8 × 100 =   ____%

9, 10, 36, 37, 54, 74, 77

Section III—Moles/Stoichiometry
☐   Number of checks ÷ 7 × 100 =   ____%

11, 12, 34, 55, 56

Section IV—Chemical Bonding
☐   Number of checks ÷ 5 × 100 =   ____%

13, 14, 15, 16, 17, 38, 39, 40, 41, 42, 43, 46, 68, 69, 70, 78, 79, 80, 81

Section V—Physical Behavior of Matter
☐   Number of checks ÷ 19 × 100 =   ____%

18, 19, 20, 44, 45, 57, 58

Section VI—Kinetics and Equilibrium
☐   Number of checks ÷ 7 × 100 =   ____%

21, 22, 23, 25, 47, 59, 60, 82, 83

Section VII—Organic Chemistry
☐   Number of checks ÷ 9 × 100 =   ____%

24, 26, 49, 71, 73

Section VIII—Oxidation-Reduction
☐   Number of checks ÷ 5 × 100 =   ____%

27, 28, 48, 50, 75, 76

Section IX—Acids, Bases, and Salts
☐   Number of checks ÷ 6 × 100 =   ____%

29, 30, 31, 32, 62, 64

Section X—Nuclear Chemistry
☐   Number of checks ÷ 6 × 100 =   ____%

Examination
August 2005

Chemistry
The Physical Setting

PART A

Answer all questions in this part.

Directions (1–30): For *each* statement or question, write in the answer space the *number* of the word or expression that, of those given, best completes the statement or answers the question. Some questions may require the use of the *Reference Tables for Physical Setting/Chemistry*.

1 Which subatomic particle has a negative charge?

    (1) proton   (3) neutron

    (2) electron   (4) positron   1 __2__

2 Which statement best describes the nucleus of an aluminum atom?

    (1) It has a charge of +13 and is surrounded by a total of 10 electrons.

    (2) It has a charge of +13 and is surrounded by a total of 13 electrons.

    (3) It has a charge of –13 and is surrounded by a total of 10 electrons.

    (4) It has a charge of –13 and is surrounded by a total of 13 electrons.   2 __2__

3 The atomic mass of an element is the weighted average of the

(1) number of protons in the isotopes of that element
(2) number of neutrons in the isotopes of that element
(3) atomic numbers of the naturally occurring isotopes of that element
(4) atomic masses of the naturally occurring isotopes of that element

3 __4__

4 In which pair do the particles have approximately the same mass?

(1) proton and electron
(2) proton and neutron
(3) neutron and electron
(4) neutron and beta particle

4 __2__

5 Two different samples decompose when heated. Only one of the samples is soluble in water. Based on this information, these two samples are

(1) both the same element
(2) two different elements
(3) both the same compound
(4) two different compounds

5 __4__

6 The elements located in the lower left corner of the Periodic Table are classified as

(1) metals   (3) metalloids
(2) nonmetals   (4) noble gases

6 __1__

7 Which of these elements has the *lowest* melting point?

(1) Li   (3) K
(2) Na   (4) Rb

7 __4__

8 Which list consists of elements that have the most similar chemical properties?

  (1) Mg, Al, and Si   (3) K, Al, and Ni

  (2) Mg, Ca, and Ba   (4) K, Ca, and Ga   8 __2__

9 The correct chemical formula for iron(II) sulfide is

  (1) FeS   (3) $FeSO_4$ Fe^2S^{-2}

  (2) Fe_2S_3 (4) $Fe_2(SO_4)_3$ FeS 9 __1__

10 Which list consists of types of chemical formulas?

  (1) atoms, ions, molecules

  (2) metals, nonmetals, metalloids

  (3) empirical, molecular, structural

  (4) synthesis, decomposition, neutralization   10 __3__

11 Which type of bonding is found in all molecular substances?

  (1) covalent bonding   (3) ionic bonding

  (2) hydrogen bonding   (4) metallic bonding   11 __1__

12 An aqueous solution of sodium chloride is best classified as a

  (1) homogeneous compound

  (2) homogeneous mixture

  (3) heterogeneous compound

  (4) heterogeneous mixture   12 __2__

13 What is the total number of electrons shared in a double covalent bond between two atoms?

  (1) 1   (3) 8

  (2) 2   (4) 4   13 __4__

14 Which formula represents a nonpolar molecule?

(1) H_2S (3) CH_4

(2) HCl   (4) NH_3 14 __3__

15 What occurs when an atom loses an electron?

(1) The atom's radius decreases and the atom becomes a negative ion.

(2) The atom's radius decreases and the atom becomes a positive ion.

(3) The atom's radius increases and the atom becomes a negative ion.

(4) The atom's radius increases and the atom becomes a positive ion.   15 __2__

16 Two samples of gold that have different temperatures are placed in contact with one another. Heat will flow spontaneously from a sample of gold at 60°C to a sample of gold that has a temperature of

(1) 50°C   (3) 70°C

(2) 60°C   (4) 80°C   16 __1__

17 Under which conditions of temperature and pressure would helium behave most like an ideal gas?

(1) 50 K and 20 kPa

(2) 50 K and 600 kPa

(3) 750 K and 20 kPa

(4) 750 K and 600 kPa   17 __3__

18 A sample of oxygen gas is sealed in container X. A sample of hydrogen gas is sealed in container Z. Both samples have the same volume, temperature, and pressure. Which statement is true?

(1) Container X contains more gas molecules than container Z.

(2) Container X contains fewer gas molecules than container Z.

(3) Containers X and Z both contain the same number of gas molecules.

(4) Containers X and Z both contain the same mass of gas.

X 18__4__

19 Which formula represents an unsaturated hydrocarbon?

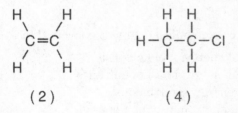

(1)

(2)

(3)

(4)

19__2__

20 Given the formula:

$$H-\underset{\underset{H}{|}}{\overset{\overset{H}{|}}{C}}-\underset{\underset{}{|}}{\overset{\overset{H}{|}}{C}}=\underset{\underset{H}{|}}{\overset{}{C}}-\underset{\underset{H}{|}}{\overset{\overset{H}{|}}{C}}-\underset{\underset{H}{|}}{\overset{\overset{H}{|}}{C}}-H$$

What is the IUPAC name of this compound?

(1) 2-pentene   (3) 2-butene

(2) 2-pentyne   (4) 2-butyne   20 __1__

21 Given the reaction system in a closed container at equilibrium and at a temperature of 298 K:

$$N_2O_4(g) \rightleftharpoons 2NO_2(g)$$

The measurable quantities of the gases at equilibrium must be

(1) decreasing   (3) equal

(2) increasing   (4) constant   21 __4__

22 Atoms of which element can bond with each other to form ring and chain structures in compounds?

(1) C   (3) H

(2) Ca   (4) Na   22 __1__

23 In a voltaic cell, chemical energy is converted to

(1) electrical energy, spontaneously

(2) electrical energy, nonspontaneously

(3) nuclear energy, spontaneously

(4) nuclear energy, nonspontaneously   23 __1__

24 In each of the four beakers shown below, a 2.0-centimeter strip of magnesium ribbon reacts with 100 milliliters of HCl(aq) under the conditions shown.

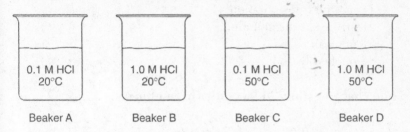

| Beaker A | Beaker B | Beaker C | Beaker D |

0.1 M HCl 20°C — Beaker A
1.0 M HCl 20°C — Beaker B
0.1 M HCl 50°C — Beaker C
1.0 M HCl 50°C — Beaker D

In which beaker will the reaction occur at the fastest rate?

(1) A   (3) C
(2) B   (4) D   24___4___

25 Which aqueous solution is the best conductor of an electrical current?

(1) 0.01 M CH_3OH (3) 0.1 M CH_3OH
(2) 0.01 M KOH   (4) 0.1 M KOH   25___4___

26 A hydrogen ion, H^+, in aqueous solution may also be written as

(1) H_2O (3) H_3O^+
(2) H_2O_2 (4) OH^- 26___3___

27 One acid-base theory states that an acid is

(1) an electron donor   (3) an H^+ donor
(2) a neutron donor   (4) an OH^- donor   27___3___

28 Which isotope will spontaneously decay and emit particles with a charge of +2?

(1) ^{53}Fe (3) ^{198}Au
(2) ^{137}Cs (4) ^{220}Fr 28___4___

29 Radioactive cobalt-60 is used in radiation therapy treatment. Cobalt-60 undergoes beta decay. This type of nuclear reaction is called

  (1) natural transmutation
  (2) artificial transmutation
  (3) nuclear fusion
  (4) nuclear fission   29 _1_

Note that question 30 has only three choices.

30 Given the balanced ionic equation:

$$2Al(s) + 3Cu^{2+}(aq) \rightarrow 2Al^{3+}(aq) + 3Cu(s)$$

Compared to the total charge of the reactants, the total charge of the products is

  (1) less
  (2) greater
  (3) the same   30 _3_

PART B–1

Answer all questions in this part.

Directions (31–50): For *each* statement or question, write in the answer space the *number* of the word or expression that, of those given, best completes the statement or answers the question. Some questions may require the use of the *Reference Tables for Physical Setting/Chemistry.*

31 The percentage by mass of Br in the compound $AlBr_3$ is closest to

(1) 10.%   (3) 75%
(2) 25%   (4) 90.%

31 __4__

32 Which symbol represents a particle with a total of 10 electrons?

(1) N   (3) Al
(2) N^{3+} (4) Al^{3+}

32 __4__

33 Which electron configuration represents an atom of aluminum in an excited state?

(1) 2-7-4   (3) 2-8-3
(2) 2-7-7   (4) 2-8-6

33 __1__

34 At STP, an element that is a brittle solid and a poor conductor of heat and electricity could have an atomic number of

(1) 12   (3) 16
(2) 13   (4) 17

34 __3__

35 Based on Reference Table S, atoms of which of these elements have the strongest attraction for the electrons in a chemical bond?

(1) Al   (3) P

(2) Si   (4) S   35 _4_

36 A sample of a compound contains 65.4 grams of zinc, 12.0 grams of carbon, and 48.0 grams of oxygen. What is the mole ratio of zinc to carbon to oxygen in this compound?

(1) 1:1:2   (3) 1:4:6

(2) 1:1:3   (4) 5:1:4   ✗36 _4_

37 Which process would most effectively separate two liquids with different molecular polarities?

(1) filtration   (3) distillation

(2) fermentation   (4) conductivity   ✗37 _1_

38 Given the balanced equation:

$$AgNO_3(aq) + NaCl(aq) \rightarrow NaNO_3(aq) + AgCl(s)$$

This reaction is classified as

(1) synthesis

(2) decomposition

(3) single replacement

(4) double replacement   ✗38 _3_

39 A solution contains 35 grams of KNO_3 dissolved in 100 grams of water at 40°C. How much *more* KNO_3 would have to be added to make it a saturated solution?

(1) 29 g   (3) 12 g

(2) 24 g   (4) 4 g   39 _1_

40 Which diagram best represents a gas in a closed container?

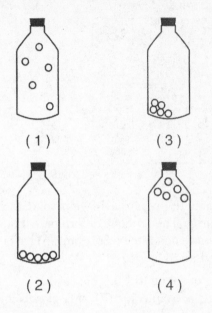

(1) (3)

(2) (4)

41 What is the total number of moles of NaCl(s) needed to make 3.0 liters of a 2.0 M NaCl solution?

(1) 1.0 mol (3) 6.0 mol
(2) 0.70 mol (4) 8.0 mol

42 Which Lewis electron-dot diagram is correct for a S^{2-} ion?

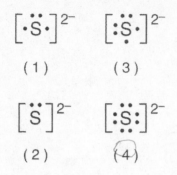

(1)   (3)

(2)   (4)   42 __3__

43 A student wants to prepare a 1.0-liter solution of a specific molarity. The student determines that the mass of the solute needs to be 30. grams. What is the proper procedure to follow?

(1) Add 30. g of solute to 1.0 L of solvent.
(2) Add 30. g of solute to 970. mL of solvent to make 1.0 L of solution.
(3) Add 1000. g of solvent to 30. g of solute.
(4) Add enough solvent to 30. g of solute to make 1.0 L of solution.   43 __4__

44 What is the total number of joules released when a 5.00-gram sample of water changes from liquid to solid at 0°C?

(1) 334 J   (3) 2260 J
(2) 1670 J   (4) 11 300 J   44 __1__

45 Which set of procedures and observations indicates a chemical change?

   (1) Ethanol is added to an empty beaker and the ethanol eventually disappears.

   (2) A solid is gently heated in a crucible and the solid slowly turns to liquid.

   (3) Large crystals are crushed with a mortar and pestle and become powder.

   (4) A cool, shiny metal is added to water in a beaker and rapid bubbling occurs.   45 ____

46 At STP, a sample of which element has the highest entropy?

   (1) $Na(s)$ (3) $Br_2(\ell)$

   (2) $Hg(\ell)$ (4) $F_2(g)$ 46 ____

47 Given the incomplete equation representing an organic addition reaction:

$$X(g) + Cl_2(g) \rightarrow XCl_2(g)$$

Which compound could be represented by X?

   (1) CH_4 (3) C_3H_8

   (2) C_2H_4 (4) C_4H_{10} 47 ____

48 Given the incomplete equation:

$$4Fe + 3O_2 \rightarrow 2X$$

Which compound is represented by X?

   (1) FeO (3) Fe_3O_2

   (2) Fe_2O_3 (4) Fe_3O_4 48 ____

49 How are $HNO_3(aq)$ and $CH_3COOH(aq)$ similar?

(1) They are Arrhenius acids and they turn blue litmus red.

(2) They are Arrhenius acids and they turn red litmus blue.

(3) They are Arrhenius bases and they turn blue litmus red.

(4) They are Arrhenius bases and they turn red litmus blue.

49 __1__

50 The chart below shows the spontaneous nuclear decay of U-238 to Th-234 to Pa-234 to U-234.

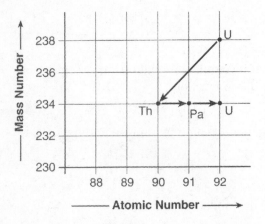

What is the correct order of nuclear decay modes for the change from U-238 to U-234?

$$^{238}_{92}U \to ^{234}_{90}Th$$

(1) β⁻ decay, γ decay, β⁻ decay

(2) β⁻ decay, β⁻ decay, α decay

(3) α decay, α decay, β⁻ decay

(4) α decay, β⁻ decay, β⁻ decay

50 __3__

PART B–2

Answer all questions in this part.

Directions (51–67): Record your answers on the answer sheet provided in the back. Some questions may require the use of the *Reference Tables for Physical Setting/Chemistry.*

51 In the space *on the answer sheet,* show a correct numerical setup for calculating the formula mass of glucose, $C_6H_{12}O_6$. [1] C: 12 XC = 72

H: Q1X12 = 12 = 180

O: 6X6 = 96

52 Write the empirical formula for the compound $C_6H_{12}O_6$. [1] CH_2O

Base your answers to questions 53 through 55 on the potential energy diagram below.

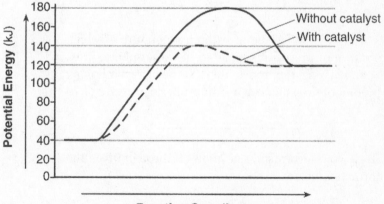

53 What is the heat of reaction for the forward reaction? [1]

140

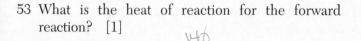

54 What is the activation energy for the forward reaction with the catalyst? [1]   *100*

55 Explain, in terms of the function of a catalyst, why the curves on the potential energy diagram for the catalyzed and uncatalyzed reactions are different. [1]   *Since a catalyst finds an alternate route lowering the activation energy*

Base your answers to questions 56 through 58 on the properties of propanone.

56 In the space *on the answer sheet*, draw the structural formula for propanone. [1]

   $-\overset{|}{\underset{|}{C}} - \overset{O}{\underset{|}{C}} - \overset{|}{\underset{|}{C}}-$

57 Explain, in terms of molecular energy, why the vapor pressure of propanone increases when its temperature increases. [1]   *Since the molecular energy increases*

58 A liquid's boiling point is the temperature at which its vapor pressure is equal to the atmospheric pressure. Using Reference Table *H*, what is the boiling point of propanone at an atmospheric pressure of 70 kPa? [1]   *45 (°C)*

Base your answers to questions 59 through 61 on the information below.

Two isotopes of potassium are K-37 and K-42.

59 What is the total number of neutrons in the nucleus of a K-37 atom? [1]   *18*

PART C

Answer all questions in this part.

Directions (68–85): Record your answers on the answer sheet provided in the back. Some questions may require the use of the *Reference Tables for Physical Setting/Chemistry.*

Base your answers to questions 68 through 70 on the information below.

> The decomposition of sodium azide, $NaN_3(s)$, is used to inflate airbags. On impact, the $NaN_3(s)$ is ignited by an electrical spark, producing $N_2(g)$ and $Na(s)$. The $N_2(g)$ inflates the airbag.

68 Balance the equation *on the answer sheet*, using the smallest whole-number coefficients. [1]

69 What is the total number of moles present in a 52.0-gram sample of $NaN_3(s)$ (gram-formula mass = 65.0 gram/mole)? [1]

70 An inflated airbag has a volume of 5.00×10^4 cm^3 at STP. The density of $N_2(g)$ at STP is 0.00125 g/cm^3. What is the total number of grams of $N_2(g)$ in the airbag? [1]

Base your answers to questions 71 through 73 on the information below.

> Element X is a solid metal that reacts with chlorine to form a water-soluble binary compound.

71 State *one* physical property of element X that makes it a good material for making pots and pans. [1]

72 Explain, in terms of particles, why an aqueous solution of the binary compound conducts an electric current. [1] *Since it is a metal(ian) its particles conduct electricity*

73 The binary compound consists of element X and chlorine in a 1:2 molar ratio. What is the oxidation number of element X in this compound? [1]

_____   *+2* $X Cl_2^{-1}$
                                          X Cl_2^{-1}

Base your answers to questions 74 through 76 on the diagram and balanced equation below, which represent the electrolysis of molten NaCl.

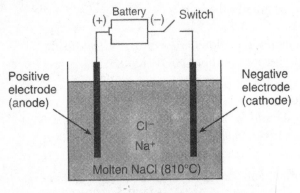

$$2NaCl \longrightarrow Cl_2 + 2Na$$

74 When the switch is closed, which electrode will attract the sodium ions? [1] *Negative*

75 What is the purpose of the battery in this electrolytic cell? [1] *to give electrical energy to complete the reaction*

76 Write the balanced half-reaction for the reduction that occurs in this electrolytic cell. [1] *$2Na^+ + 1e^- \rightarrow 2Na$*

Base your answers to questions 77 through 79 on the information below.

In a titration, 3.00 M NaOH(aq) was added to an Erlenmeyer flask containing 25.00 milliliters of HCl(aq) and three drops of phenolphthalein until one drop of the NaOH(aq) turned the solution a light-pink color. The following data were collected by a student performing this titration.

Initial NaOH(aq) buret reading: 14.45 milliliters

Final NaOH(aq) buret reading: 32.66 milliliters

77 What is the total volume of NaOH(aq) that was used in this titration?   [1]

78 In the space *on the answer sheet*, show a correct numerical setup for calculating the molarity of the HCl(aq).   [1]

79 Based on the data given, what is the correct number of significant figures that should be shown in the molarity of the HCl(aq)?   [1]

Base your answers to questions 80 through 82 on the information below.

A student was studying the pH differences in samples from two Adirondack streams. The student measured a pH of 4 in stream *A* and a pH of 6 in stream *B*.

80 Compare the hydronium ion concentration in stream *A* to the hydronium ion concentration in stream *B*.   [1]

81 What is the color of bromthymol blue in the sample from stream *A*? [1]   *yellow*

82 Identify *one* compound that could be used to neutralize the sample from stream *A*. [1] *NaOH*

Base your answers to questions 83 through 85 on the information below.

The radioisotopes carbon-14 and nitrogen-16 are present in a living organism. Carbon-14 is commonly used to date a once-living organism.

83 Complete the nuclear equation *on the answer sheet* for the decay of C-14. Include *both* the atomic number and the mass number of the missing particle. [1]

$${}^{14}_{7}N$$

84 Explain why N-16 is a poor choice for radioactive dating of a bone. [1] *Since it has a very short half life*

85 A sample of wood is found to contain ⅛ as much C-14 as is present in the wood of a living tree. What is the approximate age, in years, of this sample of wood? [1]

22860

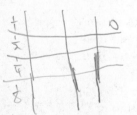

Answer Sheet
August 2005

Chemistry
The Physical Setting

PART B–2

51

52 _____

53 _____ kJ

54 _____ kJ

55 _____

_____

_____

56

57 _____

_____

_____

58 _____ °**C**

59 _____

60 _____

61 _____

_____

_____

62

63

64 and 65

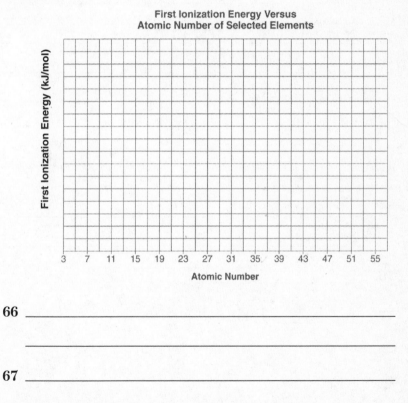

First Ionization Energy Versus
Atomic Number of Selected Elements

66 _____

_____

67 _____

_____

_____

PART C

68 __2__ NaN$_3$(s) → __2__ Na(s) → __3__ N$_2$(g)

69 _____ mol

70 _____ g

71 _____

72 _____

_____

_____

73 _____

74 _____

75 _____

_____

76 _____

77 _____ mL

78

79 _____

80 _____

_____

81 _____

82 _____

83 $^{14}_{6}C \rightarrow$ _____$^{14}_{7}N$_____ $+ ^{0}_{-1}e$

84 _____

_____

_____

85 _____ **y**

Answers
August 2005

Chemistry
The Physical Setting

Answer Key

PART A

1. 2	7. 4	13. 4	19. 2	25. 4
2. 2	8. 2	14. 3	20. 1	26. 3
3. 4	9. 1	15. 2	21. 4	27. 3
4. 2	10. 3	16. 1	22. 1	28. 4
5. 4	11. 1	17. 3	23. 1	29. 1
6. 1	12. 2	18. 3	24. 4	30. 3

PART B–1

31. 4	35. 4	39. 1	43. 4	47. 2
32. 4	36. 2	40. 1	44. 2	48. 2
33. 1	37. 3	41. 3	45. 4	49. 1
34. 3	38. 4	42. 4	46. 4	50. 4

Answers Explained

PART A

1. **2** An electron is the subatomic particle that has a negative charge.

Wrong Choices Explained:
(1), (4) A proton and a positron each have a positive charge.
(3) A neutron has no charge.

2. **2** Aluminum (Al) has an atomic number of 13; its nuclear charge is +13. In any atom, the number of protons must equal the number of electrons. Therefore, 13 electrons surround the nucleus of an aluminum atom.

3. **4** The atomic mass of an element is defined to be the weighted average of the atomic masses of the element's naturally occurring isotopes.

4. **2** The proton and neutron each have masses equal to approximately 1 atomic mass unit.

Wrong Choices Explained:
(1), (3), (4) The electron has a mass of approximately $\frac{1}{2000}$ of an atomic mass unit. A beta particle is an electron.

5. **4** Since the samples can be decomposed by a chemical change (heating), the samples are compounds. The two compounds are different because they have different physical properties (solubility in water).

6. **1** Refer to the Periodic Table of the Elements. Elements located at the left side of the table (Groups 1 and 2) are all classified as metals.

Wrong Choices Explained:
(2) Nonmetals are found in Groups 14–17.
(3) Metalloids are found in Groups 14–16.
(4) Noble gases are found in Group 18.

7. **4** See Reference Table *S*. Of the four choices given, choice (4), Rb, has the lowest melting point (312 K).

8. **2** Refer to the Periodic Table of the Elements. Elements that lie in the same group have similar chemical properties. Mg, Ca, and Ba are all located in Group 2.

9. **1** The charges of the iron (II) and sulfide ions are, respectively, 2+ and 2–. In a (neutral) compound, the sum of the charges must add to zero. Therefore, the correct formula is FeS.

10. **3** An empirical formula is one in which the atoms are given in smallest whole-number ratios. For example, the empirical formula corresponding to H_2O_2 is HO.

A molecular formula indicates the number of atoms present in a molecule, but it gives no information about the arrangement of these atoms in space or the bonds between the atoms. For example, CO_2 is a molecular formula.

A structural formula provides information about the spatial arrangement of the atoms and the bonds between the atoms. For example, O=C=O is the structural formula of CO_2.

11. **1** In all molecules, the atoms are bonded by the sharing of electrons, that is, by covalent bonding.

Wrong Choices Explained:
(2) Hydrogen bonding is an intermolecular attraction, not a bond between atoms. Not all molecular substances have hydrogen bonds.

(3), (4) Molecules do not exist in ionic or metallic substances.

12. **2** All solutions are homogeneous mixtures.

13. **4** Multiple covalent bonds share multiple pairs of electrons. In a double covalent bond, two pairs of electrons, that is, 4 electrons, are shared by two atoms.

14. **3** Nonpolar molecules must have symmetrical electron distribution. Of the choices given, only choice (3), CH_4, has a symmetrical distribution of its electrons.

15. **2** An atom has an equal number of protons and electrons. When an electron is lost, the atom becomes a positive ion. Since there is less repulsion among the outermost electrons, the radius of the ion is less than the radius of the parent atom.

16. **1** Heat will flow spontaneously between objects when a difference in temperature exists between the objects. Heat always flows from the object at the higher temperature to the object at the lower temperature.

17. **3** A "real" gas, such as helium, will behave most like an ideal gas at high temperatures and low pressures. Choice (3), 750 K and 20 kPa, is the combination with the highest temperature and the lowest pressure.

18. **3** Avogadro's principle states that equal volumes of gases at the same temperature and pressure contain equal numbers of molecules.

19. **2** An unsaturated hydrocarbon is one that contains only carbon and hydrogen and has at least one double or triple carbon-to-carbon bond. Of the choices given, only choice (2) meets these requirements.

20. **1** See Reference Tables P and Q. The hydrocarbon contains 5 carbon atoms and one double bond; its name is pentene. Since the double bonds begins on the second carbon atom, the IUPAC name is 2-pentene.

21. **4** When a reaction system reaches equilibrium, the rates of the forward and reverse reactions are equal and the concentrations of the reactants and products remain constant.

22. **1** Of the elements listed, only carbon (C) will bond with itself and form both ring and chain structures.

23. **1** By definition, a voltaic cell is an electrochemical cell that converts chemical energy to electrical energy spontaneously.

24. **4** The rate of a reaction depends on the concentration of the reactants and the temperature of the system: the more concentrated the reactant, the faster the rate; the higher the temperature, the faster the rate. Of the choices given, choice (4), Beaker D, has the highest concentration of HCl and the highest temperature.

25. **4** An electrolyte is a substance that dissolves in water and produces a solution that conducts electricity because ions are released in solution. KOH is an ionic compound that is a strong electrolyte. Choice (4), 0.1 M KOH, will be the best conductor because the concentrations of K^+ and OH^- ions are greatest.

Wrong Choices Explained:
(1), (3) CH_3OH is a molecular substance. When dissolved in water, no ions are produced in solution.

26. **3** Another way of writing $H^+(aq)$ is to show it combined with a molecule of H_2O, that is, as $H_3O^+(aq)$.

27. **3** The Brönsted-Lowry theory of acids and bases states that an acid is a proton (H^+) donor.

28. **4** See Reference Tables O and N. Alpha particles (α) are helium nuclei that have a 2+ charge. Of the choices given, only choice (4), ^{220}Fr, is an alpha emitter.

29. **1** In a transmutation reaction, one element is converted into another. When cobalt-60 undergoes beta decay, it changes to Nickel-60. Beta decay is one form of natural transmutation because it occurs spontaneously.

30. **3** The total charge on the left side of the reaction is 6+ ($3Cu^{2+}$); the total charge on the right side of the reaction is 6+ ($2Al^{3+}$).

PART B–1

31. **4** First, use the Periodic Table of the Elements and complete the following table in order to calculate the formula mass of $AlBr_3$:

Element	Atomic Mass (g/mol)	Number of Atoms in Formula	Mass of Element in Formula/g
Al	27	1	27
Br	80	3	240
		Formula mass	**267**

Second, use the formula for percent composition on Reference Table T:

$$\% \text{ composition by mass} = \frac{\text{mass of part}}{\text{mass of whole}} \times 100$$
$$= \frac{240}{267} \times 100 = \mathbf{90.\%}$$

32. **4** Refer to the Periodic Table of the Elements. The atomic number of aluminum (Al) is 13; an atom of aluminum has a total of 13 electrons. An Al^{3+} ion has three fewer electrons, or a total of 10 electrons.

33. **1** The ground state is the lowest energy state that an atom can have. When an electron moves into a higher energy level, the atom is said to be excited. Use the Periodic Table of the Elements. The ground-state configuration of aluminum (Al) is 2-8-3. Choice (1), 2-7-4, indicates that an electron in the second energy level has been "promoted" to the third energy level.

Wrong Choices Explained:
(2) This configuration has 16 electrons; it represents an excited atom of sulfur.
(4) This configuration has 16 electrons; it represents an atom of sulfur in the ground state.

34. **3** Use the Periodic Table of the Elements. The terms *brittle* and *poor conductor* describe a nonmetallic solid, such as sulfur (atomic number 16).

Wrong Choices Explained:
(1), (2) Elements 12 and 13 (Mg and Al) are metallic solids.
(4) Element 17 (Cl) is a gas at STP.

35. **4** An atom's attraction for electrons in a chemical bond is described by its electronegativity. Of the choices given, choice (4), S, has the highest electronegativity value (2.6).

36. **2** Use the formula for mole calculations on Reference Table *T*. The gram-formula mass of each element is its atomic mass, which is shown on the Periodic Table of the Elements.

$$\text{number of moles} = \frac{\text{given mass (g)}}{\text{gram-formula mass}}$$

$$\text{moles of zinc} = \frac{65.4 \text{ g}}{65.4 \text{ g/mol}} = 1.00 \text{ mol}$$

$$\text{moles of carbon} = \frac{12.0 \text{ g}}{12.0 \text{ g/mol}} = 1.00 \text{ mol}$$

$$\text{moles of oxygen} = \frac{48.0 \text{ g}}{16.0 \text{ g/mol}} = 3.00 \text{ mol}$$

$$\text{mole ratio of elements} = \mathbf{1:1:3}$$

37. **3** The process of distillation can separate two liquids on the basis of the differences in their boiling points, which depends on the differences in their molecular polarities.

Wrong Choices Explained:
(1) Filtration is generally used to separate a solid from a liquid.
(2) Fermentation is a type of organic reaction; it is not a separation technique.
(4) Conductivity is a property of a substance; it is not a separation technique.

38. **4** In a double-replacement reaction, two ionic compounds are dissolved in water. The positive and negative ions "exchange partners." One indication that a double-replacement reaction has occurred is the appearance of a precipitate (in this case, $AgCl(s)$).

Wrong Choices Explained:
(1) In a synthesis reaction, elements and/or compounds combine to produce a single compound. Example: $MgO + CO_2 \rightarrow MgCO_3$
(2) In a decomposition reaction, a single compound is broken down into a mixture of elements and/or compounds. Example: $CuS \rightarrow Cu + S$
(3) In a single-replacement reaction, a single element is exchanged with an element that is part of a compound. As a result, the previously combined element is liberated. Example: $Zn + CuSO_4 \rightarrow Cu + ZnSO_4$

39. **1** Use Reference Table G. According to the solubility curve for KNO_3, 64 grams of the substance will form a saturated solution when dissolved in 100 grams of H_2O at 40°C. Since 35 grams of the KNO_3 have already been dissolved, an additional 29 grams is needed to saturate the solution.

40. **1** Gases consist of particles that are free to move randomly and fill the entire container. Of the choices given, only the container in choice (1) illustrates these properties.

Wrong Choices Explained:
(2), (3) These containers appear to have solids in them.
(4) This container illustrates a situation that is not possible.

41. **3** Use the concentration (molarity) formula given on Reference Table T:

$$\text{molarity} = \frac{\text{moles of solute}}{\text{liters of solution}}$$
$$\text{moles of solute} = \text{molarity} \cdot \text{liters of solution}$$
$$= (2.0 \text{ M}) \cdot (3.0 \text{ L}) = \textbf{6.0 mol}$$

42. **4** Use the Periodic Table of the Elements. An atom of sulfur (S) contains 6 valence electrons. An S^{2-} ion contains 8 electrons arranged in four pairs as illustrated in choice (4).

43. **4** Since the volume of the solvent is not known, the solvent must be added to the 30. g of solid until a total volume of 1.0 L of solution is obtained.

44. **2** Use Reference Table B (heat of fusion) and the heat formula on Reference Table T:

$$q = m H_f$$
$$= (5.00 \text{ g}) \cdot (334 \text{ J}/\text{g}) = \textbf{1670 J}$$

45. **4** When a chemical change occurs, a chemical reaction takes place and one or more new substances are formed. The description given in choice (4) indicates that a gas has been produced—evidence that a chemical reaction has occurred.

Wrong Choices Explained:
(1), (2), (3) Each of these choices—evaporation, melting, and crushing—are examples of physical changes. No new substances have been produced.

46. **4** Entropy is a measure of disorder. A substance in the gas phase has the highest degree of disorder, while a solid has the lowest degree of disorder.

47. **2** Alkenes and alkynes (hydrocarbons containing double or triple carbon-to-carbon bonds, respectively) can undergo addition reactions. See Reference Table Q. Of the choices listed, only choice (2), C_2H_4, is an alkene.

Wrong Choices Explained:
(1), (3), (4) These hydrocarbons are alkanes. They contain no double or triple bonds.

48. **2** There are a total of 4 Fe atoms and 6 O atoms on the left side of the equation. Since the product is $2X$, the formula unit X must be Fe_2O_3.

49. **1** See Reference Tables K and M. HNO_3 and CH_3COOH are classified as Arrhenius acids because both can liberate H^+ ions in aqueous solution. The color of litmus in an acidic solution is red.

50. **4** Complete each nuclear equation, and use Reference Table O to identify the decay products:

$$^{238}_{92}U \rightarrow \; ^{234}_{90}Th + ^4_2He$$

$$^{234}_{90}Th \rightarrow \; ^{234}_{91}Pa + ^0_{-1}e$$

$$^{234}_{91}Pa \rightarrow \; ^{234}_{92}U + ^0_{-1}e$$

^4_2He is an alpha (α) particle; $^0_{-1}e$ is a beta (β^-) particle.

The sequence is α decay, β^- decay, β^- decay.

PART B–2

[Point values are indicated in brackets.]

51. Use the following table and the Periodic Table of the Elements:

Element	Atomic Mass (g/mol)	Number of Atoms in Formula	Mass of Element in Formula/g
C	12.0	6	$(12.0)\cdot(6)$
H	1.0	12	$(1.0)\cdot(12)$
O	16.0	6	$(16.0)\cdot(6)$
		Formula mass	$\mathbf{(12.0)\cdot(6) + (1.0)\cdot(12) + (16.0)\cdot(6)}$

[1 point]

52. An empirical formula is one in which the elements are expressed in simplest whole-number ratios. Divide the formula $C_6H_{12}O_6$ to obtain **CH_2O**. [1 point]

53. Refer to the diagram accompanying this question. The heat of reaction is the difference between the potential energy of the reactants, located at 40 kJ, and the potential energy of the products, located at 120 kJ:

$$\Delta H = 120 \text{ kJ} - 40 \text{ kJ} = \textbf{+80 kJ}$$

[1 point]

54. Refer to the diagram accompanying this question. The activation energy of the forward catalyzed reaction is the difference between the potential energy of the reactants and the peak of the dotted curve, located at 140 kJ:

$$E_a = 140 \text{ kJ} - 40 \text{ kJ} = \textbf{+100 kJ}$$

[1 point]

55. A catalyst provides an alternate pathway for the reaction. This pathway has a lower activation energy than that of the uncatalyzed reaction. [1 point]

56. Use Reference Tables *P* and *R*. The compound propanone has the prefix prop-, which means it has three carbon atoms. It also has the suffix –one, which means that it is a ketone, that is, the carbonyl group (C=O) is located on the second carbon atom. The structure is shown below:

$$
\begin{array}{ccc}
\text{H} & \text{O} & \text{H} \\
| & \| & | \\
\text{H–C–C–C–H} \\
| & & | \\
\text{H} & & \text{H}
\end{array}
$$

[1 point]

57. As the temperature increases, the average kinetic energy of the propanone molecules increases. The increased kinetic energies allow more propanone molecules to escape the liquid phase. [1 point]

58. Use Reference Table *H* to determine the temperature at which the vapor pressure curve for propanone reaches the vapor pressure line of 70 kPa.

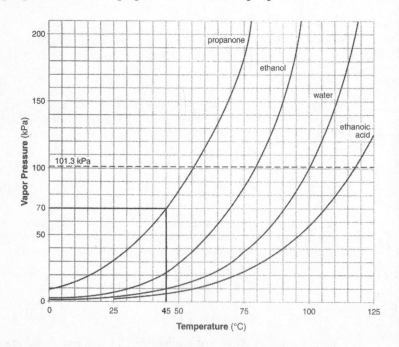

The calculated boiling point of propanone is **45°C ± 2°C** [1 point]

59. Use the Periodic Table of the Elements. The atomic number of potassium (K) is 19; there are 19 protons in the nucleus of each potassium atom. The mass number of this isotope is 37; there are a total of 37 protons and neutrons in the nucleus of this isotope. Subtract the atomic number from the mass number to obtain the number of neutrons:

$$37 \text{ protons and neutrons} - 19 \text{ protons} = \textbf{18 neutrons}$$

[1 point]

60. The valence electrons are those electrons that occupy the outermost level of the atom. Use the Periodic Table of the Elements. The electron configuration of potassium (K) is 2-8-8-1; an atom of potassium has **1 valence electron**. [1 point]

61. K-37 and K-42 are both isotopes of potassium because they have the same number of protons (19) but different numbers of neutrons (18 and 23, respectively). [1 point]

62. Use the combined gas law formula on Reference Table T:

$$\frac{P_1 V_1}{T_1} = \frac{P_2 V_2}{T_2}$$

$$V_2 = \frac{P_1 V_1 T_2}{T_1 P_2} = \frac{\left(101.3\,\text{kPa}\right)\cdot\left(20.0\,\text{mL}\right)\cdot\left(283\,\text{K}\right)}{\left(297\,\text{K}\right)\cdot\left(94.6\,\text{kPa}\right)}$$

[1 point]

63. In the Lewis electron-dot diagram for PCl_3, P is the central atom and it shares a single pair of electrons with each chlorine atom. Every atom in the molecule has a total of 8 electrons. See the diagram below:

$$:\!\ddot{C}l\!:$$
$$|$$
$$:\!\ddot{C}l\!-\!P\!-\!\ddot{C}l\!:$$

[1 point]

64–67. Refer to the diagram below:

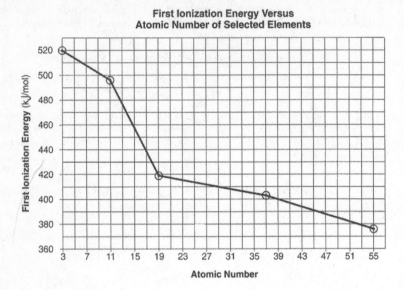

**First Ionization Energy Versus
Atomic Number of Selected Elements**

64. One point is awarded for providing an appropriate scale on the y-axis. An appropriate scale is one that allows a trend to be observed. [1 point]

65. One point is awarded for plotting all of the data points correctly (±0.3 of a grid space). [1 point]

66. Refer to the graph that appears immediately before question 64. As the atomic number increases, the first ionization energy decreases. [1 point]

67. Of the elements given in the table corresponding to questions 64–67, cesium has a larger atomic radius than rubidium (see Reference Table S). The larger size of the atom results in a smaller force of attraction between the nucleus and the valence electron. Therefore, removing a valence electron from cesium is easier than from rubidium. This is confirmed by the fact that cesium has a lower first ionization energy than rubidium. [1 point]

PART C

[Point values are indicated in brackets.]

68. The balanced equation, using smallest whole-number coefficients is:

$$2NaN_3(s) \rightarrow 2Na(s) + 3N_2(g)$$

[1 point]

69. Use the mole calculation formula on Reference Table T:

$$\text{number of moles} = \frac{\text{given mass (g)}}{\text{gram-formula mass}}$$

$$= \frac{52.0 \text{ g}}{65.0 \text{ g/mol}} = \textbf{0.800 mol}$$

[1 point]

70. Use the density formula on Reference Table T:

$$d = \frac{m}{V}$$

$$m = d \cdot V$$

$$= \left(0.00125 \text{ g/cm}^3\right) \cdot \left(5.00 \cdot 10^4 \text{ cm}^3\right) = \textbf{62.5 g}$$

[1 point]

71. Acceptable responses include:
   - The metal conducts heat well.
   - The metal is malleable.
   - The metal has a high melting point.

[1 point]

72. An aqueous solution conducts an electric current because of the presence of *mobile ions* in the solution. [1 point]

73. Use the Periodic Table of the Elements. The formula of the compound is XCl_2. The (negative) oxidation number of the chloride ion is -1. Since the compound is neutral, the sum of the oxidation numbers must equal 0:

$$x + 2(-1) = 0$$

$$x = +2$$

[1 point]

74. Sodium ions are positive, and they will be attracted to the negative electrode. [1 point]

75. The battery provides the energy needed for the nonspontaneous reaction (electrolysis of molten NaCl) to occur. [1 point]

76. Reduction involves a gain of electrons. Na^+ gains an electron to produce Na. Either of the two half-reactions given below is acceptable:

$$Na^+ + e^- \rightarrow Na$$

$$2Na^+ + 2e^- \rightarrow 2Na$$

[1 point]

77. Subtract the initial buret reading from the final buret reading:

$$32.66 \text{ mL} - 14.45 \text{ mL} = \textbf{18.21 mL } NaOH(aq)$$

[1 point]

78. Use the titration formula on Reference Table T:

$$M_A V_A = M_B V_B$$

$$M_A = \frac{M_B V_B}{V_A}$$

$$= \frac{(3.00 \text{ M}) \cdot (18.21 \text{ mL})}{(25.00 \text{ mL})}$$

[1 point]

79. The answer must contain the same number of significant figures as the data value with the *smallest* number of significant figures. Since the molarity of the NaOH contains only 3 significant figures, the answer must also contain only **3 significant figures**. [1 point]

80. The pH is inversely related to the number of hydronium ions in a solution. Since stream A has a lower pH, the concentration of hydronium ions in this stream is higher (by a factor of 100). [1 point]

81. Use Reference Table M. The color of bromthymol blue is **yellow** at any pH less than 6.0. [1 point]

82. Since the sample in stream A is acidic, any base (such as NaOH or NH_3) could be used to neutralize the sample. [1 point]

83. In balancing a nuclear equation, the upper and lower sets of numbers on both sides of the equation must be equal:

$$^{14}_{6}C \rightarrow {}^{14}_{7}N + {}^{0}_{-1}e$$

See the Periodic Table of the Elements. Since the daughter nucleus has an atomic number of 7, it must be nitrogen (N). [1 point]

84. Use Reference Table N. The half-life of N-16 is only 7.2 s; it decays much too rapidly for it to be of use in dating a sample of bone. [1 point]

85. Use Reference Table N. The half-life of C-14 is 5730 y. Use the radioactive decay formula on Reference Table T:

$$\text{fraction remaining} = \left(\frac{1}{2}\right)^{\frac{t}{T}}$$

$$\frac{1}{8} = \left(\frac{1}{2}\right)^3 = \left(\frac{1}{2}\right)^{\frac{t}{T}}$$

$$\frac{t}{T} = 3$$

$$t = 3T = (3) \cdot (5730 \text{ y}) = \textbf{17,190 y}$$

[1 point]

Mark (✓) the questions you answered correctly. Count the number of checks and follow the formulas given to determine your score on each topic.

Core Area	☐ Questions Answered Correctly

8, 58, 64, 65, 66, 79

Section M—Math Skills
☐ Number of checks ÷ 6 × 100 = ___03___ %

1, 2, 3, 4, 32, 33, 60

Section I—Atomic Concepts
☐ Number of checks ÷ 7 × 100 = _____ %

6, 7, 8, 34, 59, 61, 67, 70, 71

Section II—Periodic Table
☐ Number of checks ÷ 9 × 100 = _____ %

5, 9, 10, 30, 31, 36, 38, 48, 51, 52, 68, 69

Section III—Moles/Stoichiometry
☐ Number of checks ÷ 12 × 100 = _____ %

11, 13, 14, 15, 35, 42, 63

Section IV—Chemical Bonding
☐ Number of checks ÷ 7 × 100 = _____ %

12, 16, 17, 18, 37, 39, 40, 41, 43, 44, 45, 57, 62

Section V—Physical Behavior of Matter
☐ Number of checks ÷ 13 × 100 = _____ %

21, 24, 46, 53, 54, 55

Section VI—Kinetics and Equilibrium
☐ Number of checks ÷ 6 × 100 = _____ %

19, 20, 22, 47, 56

Section VII—Organic Chemistry
☐ Number of checks ÷ 5 × 100 = _____ %

23, 73, 74, 75, 76

Section VIII—Oxidation-Reduction
☐ Number of checks ÷ 5 × 100 = _____ %

25, 26, 27, 49, 72, 77, 78, 80, 81, 82

Section IX—Acids, Bases, and Salts
☐ Number of checks ÷ 10 × 100 = _____ %

28, 29, 50, 83, 84, 85

Section X—Nuclear Chemistry
Number of checks ÷ 6 × 100 = ___100___ %